Indian Livestock Breeds

NIPA® GENX ELECTRONIC RESOURCES & SOLUTIONS P. LTD.
New Delhi-110 034

About the Authors

Dr. Biswanath Patra, currently, Professor and Head in Animal Genetics & Breeding at MB Veterinary College, Dungarpur, Rajasthan. He has also worked as Associate Professor and Head in Animal Genetics & Breeding at RPS Veterinary College, Mahendragarh, affiliated with LUVAS, Hissar. He has more than twelve years of working experience with various aspects in Molecular Genetics, Pharmacology, Immunogenetics, Bioinformatics and Biostatistics in USA with several high impact NCBI publications including nature group journal. He has also teaching experiences in Animal Genetics & Breeding and Biostatistics. He did Master's in Genetics and Animal Breeding from GBPUA&T, Pantnagar and a Ph.D. in Animal Breeding from Indian Veterinary Research Institute (IVRI).

He did Postdoctoral Research in several world famous institutes including UT South-western Medical Centre at Dallas, the home of seven noble laureates till date. He learned genome Bioinformatics at Thomas Jefferson University, Philadelphia. He has given several awards wining oral and poster presentations from his research works in national and international symposiums. For several years he has presented his research works at highly reputed Experimental Biology meetings in USA. E-Email:patrabn1@gmail.com, biswanathpatra1@gmail.com, Phone no. +91-8777806047(M).

Dr. Bharat Bhushan, PhD, ARS 1985 currently working as Professor, MJF College of Veterinary and Animal Sciences, Chomu, Jaipur, Rajasthan. Before joining to this new position he has superannuated as Principal Scientist & Head, Division of Animal Genetics, ICAR-Indian Veterinary Research Institute (ICAR-IVRI), Izatnagar – 243 122 (UP), India. He did his PG and PhD in Animal Breeding from G. B. Pant University of Agriculture & Technology, Pantnagar, India. He was selected for Agricultural Research Service (ARS) in the year 1985 and joined as Scientist in the discipline of Animal Genetics & Breeding at ICAR. He had more than 38 years of experience in conducting research on various aspects of Animal Genetics, Animal Breeding, Molecular Genetics and Genomics. Published more than 261 research papers in journals of national and international repute, co-authored 9 lead/review papers, 150 invited lectures and presented more than 350 research papers in various national and international conferences/symposia. He was involved in guidance of more than 100 research scholars for their MVSc and PhD degrees (30 as Chairman: 16 MVSc; 14 PhD). He had undergone the Advanced training on "Genetic resistance to diseases", Washington State University, Pullman, USA. Recipient of several national awards/recognition including the Best Teacher Award 2006-07 of ICAR-IVRI and Best Paper Award for highest Impact Factor by ICAR-IVRI. As Program Coordinator, successfully organized an International Training for Afghan Scientists sponsored by Ministry of External Affairs, Government of India, New Delhi. Patented a portable large animal restraining device "TRAVIPORT" Design No. 255486. Evaluated a number of Masters and PhD thesis received from various Institutions/Universities of India. Also evaluated a PhD thesis received from Charles Sturt University, Locked Bag 588, Boorooma St., Wagga Wagga, NSW, Australia.

Indian Livestock Breeds

Biswanath Patra
Professor and Head
Animal Genetics & Breeding
MB Veterinary College, Dungarpur-314001, Rajasthan
B.V.Sc. & A.H. (Kolkata), M.V.Sc. (GBPUA&T, Pantnagar)
Ph.D. (IVRI, Izatnagar, UP)
Postdoctoral Fellow, USA (2005 to 2016), UT Southwestern Medical Center at Dallas, Thomas Jefferson University, Philadelphia, LLUMC California
UAMS Little Rock
Associate Professor and Head in Animal Genetics & Breeding
RPS College of Veterinary Sciences, Balana

Bharat Bhushan, Ph.D., ARS 1985
Professor, Animal Genetics and Breeding
MJF College of Veterinary and Animal Sciences
Chomu, Jaipur, Pin-301408 (Rajasthan)
Former Principal Scientist & Head, Division of Animal Genetics
Visiting Scientist WSU, USA
ICAR-Indian Veterinary Research Institute
Izatnagar - 243 122, Bareilly, Uttar Pradesh, India

NIPA® GENX ELECTRONIC RESOURCES & SOLUTIONS P. LTD.
New Delhi-110 034

**NIPA® GENX ELECTRONIC
RESOURCES & SOLUTIONS P. LTD.**

101,103, Vikas Surya Plaza, CU Block
L.S.C.Market, Pitam Pura, New Delhi-110 034
Ph. +91 11 27341616, 27341717, 27341718
E-mail: newindiapublishingagency@gmail.com
www: www.nipabooks.com

For customer assistance, please contact
Phone: + 91-11-27 34 17 17
Fax: + 91-11-27 34 16 16
E-Mail: feedbacks@nipabooks.com

Print ISBN: 978-93-5887-273-6

ebook ISBN: 978-93-5887-769-4

Composed and Designed by NIPA®.

Preface

The book "Indian Livestock Breeds" has been written for the students and researchers of veterinary and animal sciences with latest information so as to help them to learn about the livestock breeds and to apply the appropriate knowledge and description. All the chapters in the book cover basic topics at Undergraduates and Postgraduates levels at Indian universities and colleges as per VCI syllabus.

India has its own dairy cattle breeds like one of the most important milch breeds is Gir, which is known for its robustness and high milk yield. The Gir breed of cattle hails from Gujarat and originated in Gir forest. Another best milch breed is Sahiwal, originated from Punjab. Sahiwal cattle are prized for their heat tolerance as well for milk production characteristics. Besides, the Red Sindhi is also known for high milk production. The animals of this breed are well-adapted to hot and harsh climatic conditions. The animals of this breed are found in Sindh now in Pakistan and in Rajasthan. Tharparkar is also categorised into a dairy breed, which is resilient and drought-resistant, and thrive well in arid regions. The Indian economy is also greatly influenced by buffalo farming, as they contribute 56% to the total milk production to India. Mostly swamp buffaloes found permanently in marshy areas, where they wallow in mud and eat coarse marsh grasses. The most popular breeds of buffalo in India are Murrah, Nili-Ravi, Jaffarabadi, Surti, Mehsana, Kundi, Nagpuri and Bhadawari etc. The Banni buffalo breed is highly adapted to local climatic conditions in the areas and traditionally they are reared under the extensive methods of night grazing.

Indian Council of Agricultural Research- National Bureau of Animal Genetic Resources (ICAR-NBAGR), Karnal (Haryana) has registered seven new breeds of indigenous livestock species and one synthetic cattle breed in the country (5th December, 2023). These breeds are Aravali Chicken (Gujarat), Andamani Duck (Andaman and Nicobar Islands), Anjori Goat (Chhattisgarh); Andamani Goat (Andaman and Nicobar Islands); Bhimthadi Horse (Maharashtra), Andamani Pig (Andaman and Nicobar Islands), Macherla Sheep (Andhra Pradesh), and Frieswal Cattle (Uttar Pradesh and Uttarakhand). After inclusion of these breeds, total number of registered breeds has been reached to 220, including 53 for cattle, 20 for buffaloes, 39 for goats, 45 for sheep, 8 for horses & ponies, 9 for camel, 14 for pigs, 3 for donkeys, 3 for dogs, 1 for yak, 20 for chicken, 3 for ducks, 1 for geese and 1 for synthetic cattle. ICAR-NBAGR also allotted Accession numbers to these newly registered breeds.

The book chapters have been framed according to the syllabus of Veterinary Council of India (VCI). The syllabus of BVSc&AH, BSc (Ag), PG and PhD students in Animal Genetics & Breeding and Livestock Production & Management syllabus has been covered in 14 chapters, which will help students to practice and prepared for various examinations conducted by universities and ICAR like JRF, SRF, ARS and CSIR etc. All the Veterinary Colleges , Agriculture Colleges in India and also Animal Science Colleges all over the world teach about the domestic animal breed characteristics to the students of Animal Genetics & Breeding during UG, PG and PhD levels with the aim to learn and develop the concept for livestock improvement, which is an integrated component of agriculture over the globe Rearing of improved breed of livestock by the farmers leads to sustainable growth of the nation as well as farming community as a whole which is the need of current time.

Besides, the Indian Livestock Breeds book has a lot of important attractive features such as easy to understand, compiled and arranged within simple English language. All the breeds have been represented with photographs and detailed descriptions which will be very helpful for UG and PG students. The authors have provided a lot of information in the book for better understanding about the subject and for the preparation of competitive examinations. We have also included question answer sets in the book to help the students for VCI based UG and PG levels competitive examinations.

Authors

Acknowledgement

My sincere thanks to Mr. Ashish Anand (Director), Mr. Kapil Anand (Secretary), Mr. Yashwant Anand (Financial Secretary) and Dr. Vinod Kumar Sharma (Dean) of MB Veterinary College Dungarpur for motivating to write this book for the college students.

It is my great pleasure to express my indebts to my parents Mr. Ashutosh Patra (retired high school teacher in science) who teaches me in school life and my mother Mrs. Basanti Patra for their sacrifice and encouragement for pursuing my academic career and writing helpful books for students. I express my sincere gratitude to beloved Gurudev, Lord Shiva and Maa Durga for blessing me with my second life to write this book. I am also highly thankful to Pratima Kuity Patra, Shri Dulal Kuity and Mrs. Gauri Kuity for continuing inspiration to my academic pursuits.

I express my special thanks to my Postdoctoral mentors Dr. Abbas Parsian (UAMS), Dr. David Allen Boothman (UTSW), Dr. Rajanikanth Vadigepalli (TJU) and Dr. David Baylink (LLUMC). I also like to acknowledged my close friends Dr. Vikram Arya (MBBS) and Dr. Samar Biswas (MBBS), Dr. Mohammad Karim (MBBS), Dr. Swapan Kumar Das, Dr. Bijay Kumar Behera and Dr. Bala Manickam for theirs support during academic careers.

I wish to express my gratitude to everyone for helping to publish this book. I extend my special acknowledgement to Devi Prasad Isore, Indranil Samata, Parth Gaur, Hitesh Purohit, Shashikant Choudhury and Ajay Saha for their encouragement and assistance during manuscript writing, special thanks to Dr RKS Bais Principal Scientist, Division of Poultry Genetics & Breeding, ICAR-Central Avian Research Institute, (ICAR-CARI)(retired), Izatnagar, Bareilly, India and Dr Bharat Bhushan, Principal Scientist & Head(retired), Division of Animal Genetics, ICAR-Indian Veterinary Research Institute, (ICAR-IVRI), Izatnagar, Bareilly, India who guided me during M.V. Sc and Ph.D. in the discipline in the Animal Genetics & Breeding, respectively.

I voiced my deep sense of appreciation to MB Veterinary College Dungarpur, Rajasthan for offering me Head and Professor rank positions in Animal Genetics and Breeding which highly inspired me to write this book for UG and PG levels of students for a better concept about the Livestock breeds. Authors will be very happy if any student or reader communicates regarding the basic mistakes or suggestions for its further improvement in the next edition.

Authors

Contents

1

Cattle Breeds of India and the World

Important features

- Cattle produce 81% of world milk production, followed by buffaloes with 15 percent, goats with 2 percent and sheep with 1 percent; camels provide 0.5 percent. The remaining share is produced by other dairy species such as equines and yaks.
- About one-third of milk production in developing countries comes from buffaloes, goats, camels and sheep. In developed countries, almost all milk is produced by cattle.
- Milk from dairy species other than cattle represents 40 percent of milk production in Asia, 23 percent in Africa, 3 percent in Europe and 0.5 percent in the Americas; it is almost non-existent in Oceania (FAO).

Taxonomy of Cattle

Cattle belong to the kingdom Animalia, phylum Chordata, class Mammalia, order *Artiodactyla*, family *Bovidae*, genus *Bos*, and species *taurus* or *indicus*. Cattle were originally identified by Carolus Linnaeus as three separate species: *Bos taurus, Bos indicus, and Bos primigenius*. Bovines, which include cattle, bison, African buffalo, water buffalos, and certain antelopes, belong to the subfamily *Bovinae*.

Scientific classification

Domain:	Eukaryota
Kingdom:	Animalia
Phylum:	Chordata
Class:	Mammalia
Order:	Artiodactyla
Family:	*Bovidae*

Subfamily: *Bovinae*

Genus: *Bos*

Species: *B. taurus B. indicus*

Binomial name: *Bos taurus, Bos indicus.*

Bovines (subfamily *Bovinae*) comprise a diverse group of 10 genera of medium to large-sized ungulates, including cattle, bison, African buffalo, water buffalos, and the four-horned and spiral-horned antelopes.

Genus Bos

Aurochs, *Bos primigenius*

Eurasian aurochs, *B. p. primigenius*

Indian aurochs, *B. p. namadicus*

Banteng, *Bos javanicus*

Gaur, *Bos gaurus*

Gayal, *Bos frontalis*

Domestic yak, *Bos grunniens*

Wild yak, *Bos mutus*

Bos palaesondaicus,

Kouprey, *Bos sauveli* (possibly extinct)

Domestic cattle, *Bos taurus*

Taurine cattle, *B. taurus*

Zebu cattle, *B. indicus*

Sanga cattle, *B. africanus*

Cattle are descended from a wild ancestor called the **aurochs**. The aurochs were huge animals that originated on the subcontinent of India and then spread into China, the Middle East, and eventually northern Africa and Europe. Aurochs are one of the animals painted on the famous cave walls near Lascaux, France. People started domesticating aurochs between 8,000 and 10,000 years ago. Cattle were domesticated after sheep, goats, pigs, and dogs.

Cattle were first brought to the western hemisphere by Columbus on his second voyage to the New World in 1493. Spanish explorer Hernando Cortez took the offspring of the same cattle to Mexico in 1519. Juan Bautista de Anza (1773) brought 200 heads of cattle to California to supply the early California missions.

There are two classifications: ***Bos indicus*** and ***Bos taurus***. ***Bos taurus*** includes British and Continental. British breeds, also known as English breeds, are smaller in size than Continental breeds. These breeds are the foundation of the U.S. beef herds. Common English breeds include Angus, Red Angus, Shorthorn and Hereford. Continental breeds, also called Exotics, originated in Europe. These are larger, lean, muscular and can tolerate hot climates. Continental breeds include Charolais, Limousin, Simmental and Salers.

Bos indicus are humped cattle originating from South Central Asia. They are adapted to the stresses of heat, humidity, parasites, and poorly digestible forages. *Bos indicus* breeds are often found in the southern United States. Common *Bos indicus* breeds are Brahman, Brangus, Beefmaster, Simbrah and Santa Gertrudis.

Important notes

- In developed countries cow milk production is decreasing, together with the numbers of dairy operations and animals, but productivity per cow is increasing. In developing countries, production is increasing, together with the number of lactating cows.
- **Cattle produce about three-quarters of milk production in sub-Saharan Africa, about 60 percent in Asia and nearly all the milk produced in Latin America (FAO).**
- Average milk yields vary widely among countries, mainly because of differences in production systems (e.g., animal nutrition, breeds). In countries such as Bangladesh and Nigeria, the average cattle milk yield is ≤ 500 kg/year. In countries with developing dairy sectors, such as the Islamic Republic of Iran, Peru and Vietnam, the average cattle milk yield is > 2 000 kg/year.
- In much of Asia milk is becoming the main output of cattle production.
- **Major producers of cow milk are India, the United States of America and Brazil.**
- The Holstein-Friesian is the most widespread cattle breed in the world; it is present in more than 150 countries.
- Specialized dairy breeds (*Bos taurus*) are almost exclusively used in temperate and developed regions; most of the cattle in developing

countries, particularly in the humid tropics, are of the zebu type (*Bos indicus*).

- The countries with the most dairy cattle are India, Brazil, China and Pakistan.

How many breeds of cattle are identified in India?

Our country has vast genomic biodiversity; after including all breeds, the total number of registered indigenous cattle breeds are 53 and including Alambadi it is 54.

The synthetic dairy cattle namely **Frieswal**, a newly registered breed, which is one of the crossbred of Holstein Friesian (62.5) and Sahiwal (37.5) inheritance, is capable of producing about 7,000 kg of milk yield in a standard lactation with a peak yield of about 41 kg. As per the reports this breed is acclimatised to all agro-climatic regions of the country. India has the highest cattle population (192.49 million) in the world producing 24%, and the highest milk yield (221.1 million tons) in the world.

Karyotype of cattle

- The karyotype of cattle typically consists of 58 acrocentric autosomes (non-sex chromosomes) and 2 sexual chromosomes.
- C-banding revealed C-positive heterochromatin in centromeric regions almost in all chromosomes.
- The sexual chromosomes include a submetacentric X chromosome and a Y chromosome. The morphology of the Y chromosome varies depending on the cattle sub-species.
- **Specifically, the Y-chromosome found in *Bos taurus* (European cattle) is submetacentric, while that in *Bos indicus* (Indian cattle) is small acrocentric.**

Robertsonian translocation in cattle

- The Robertsonian translocation is a structural chromosomal abnormality that involves the fusion of two acrocentric chromosomes at or around the centromere, resulting in the formation of a metacentric chromosome.
- In this type of translocation, **the number of chromosomes is reduced by combining two separate chromosomes into one**. Specifically, the chromosomes that combine in these defects are the largest and smallest ones, numbered 1 and 29. Hence, it is called a 1:29 Robertsonian translocation.

- The Robertsonian 1/29 translocation is the most frequent structural chromosomal abnormality in cattle.
- **The first description of 1/29 originated from the Swedish Red and White cattle breed (SRB) in 1964 by Ingemar Gustavsson.**
- It has been documented in various frequencies in about 60 different breeds of both *Bos taurus* (European cattle) and *Bos indicus* (Indian cattle).
- Effect on Fertility: Robertsonian translocations have an adverse effect on fertility.

State wise cattle breeds in India

Sl. No.	Breeds	Breeding tract	Utility
1	Ongole	Andhra Pradesh	Milk and Draught
2	Punganur	Andhra Pradesh	Milk and Draught
3	Lakhimi	Asam	Milk and Draught
4	Bachaur	Bihar	Draught
5	Purnea	Bihar	Milk and Draught
6	Kosali	Chattisgarh	Draught
7	Shweta Kapila	Goa, North and South	Milk
8	Dagri	Gujarat	Draught
9	Kankrej	Gujarat and Rajasthan	Milk and Draught
10	Gir	Gujarat	Milk
11	Belahi	Haryana and Chandigarh	Milk and Draught
12	Hariana	Haryana,	Milk and Draught
13	Himachali Pahari	Himachal Pradesh	Milk and Draught
14	Ladakhi	J &K	Draught
15	Amritmahal	Karnataka	Transport and Draught
16	Hallikar	Karnataka	Draught
17	Malnad Gidda	Karnataka	Draught
18	Krishna Valley	Karnataka and Maharashtra	Draught
19	Vechur	Kerala	Milk and Manure
20	Malvi	Madhya Pradesh	Draught
21	Nimari	Madhya Pradesh	Draught
22	Red Kandhari	Maharashtra	Draught
23	Kokan Kapila	Maharashtra	Draught
24	Khillar	Maharashtra and Karnataka	Draught
25	Kathani	Vidarbha region of Western Maharashtra.	Dual purpose
26	Gaolao	Maharashtra and Madhya Pradesh	Milk and Draught
27	Dangi	Maharashtra and Gujarat	Draught

Sl. No.	Breeds	Breeding tract	Utility
28	Deoni	Maharashtra and Karnataka	Milk and Draught
29	Masilum	Meghalaya	Sports, milk, manure and socio-cultural festivals
30	Thutho	Nagaland	Draught and Meat
31	Binjharpuri	Odisha	Milk and Draught
32	Khariar	Odisha	Draught
33	Motu	Odisha	Draught
34	Ghumusari	Odisha	Draught
35	Sahiwal	Punjab and Rajasthan	Milk
36	Red Sindhi	Punjab, Orissa, Tamil Nadu, Bihar, Kerala and Assam	Milk
37	Mewati	Rajasthan, Haryana and Uttar Pradesh	Draught
38	Nagori	Rajasthan	Draught
39	Nari	Rajasthan and Gujarat	Milk and Draught
40	Rathi	Rajasthan	Milk
41	Sanchori	Rajasthan	Milk and manure
42	Tharparkar	Rajasthan and Gujarat	Milk and Draught
43	Siri	Sikkim and West Bengal	Draught
44	Umblachery	Tamil Nadu	Draught
45	Pulikulam	Tamil Nadu	Draught and Game (Jallikattu)
46	Kangayam	Tamil Nadu	Draught
47	Bargur	Tamil Nadu	Draught
48	Alambadi	Alambadi on the bank of river Cauvery, Tamil Nadu	Draught
49	Poda Thurpu	Telangana	Milk and Draught
50	Gangatiri	Uttar Pradesh and Bihar	Milk and Draught
51	Kenkatha	Uttar Pradesh and Madhya Pradesh	Draught
52	Kherigarh	Uttar Pradesh	Draught
53	Ponwar	Uttar Pradesh	Draught
54	Badri	Uttarakhand	Milk and Draught

Recently registered breeds of cattle by ICAR-National Bureau of Animal Genetics Resources (NBAGAR), Karnal, (Haryana)

Cattle registered in year 2016

1. Badri
2. Lakhami

Cattle registered in year 2018

1. Ladakhi
2. Konkan Kapila

Cattle registered in year 2020

1. Purnea
2. Nari
3. Dagri
4. Himanchali Pahari
5. Shweta Kapila
6. Thutho
7. Poda Thurpu

Cattle registered in year 2022

1. Kathani
2. Masilum
3. Sanchori

Cattle registered in year 2023

1. Frieswal

Indigenous Breeds

Indigenous breeds are classified under three groups based on utility / purpose.

a) Milch breeds / Milk breeds

b) Dual purpose breeds

c) Draught breeds

a) Milch breeds / Milk breeds

The cows of these breeds have high milk yields and the male animals are slow or poor work animals. The examples of Indian milch breeds are **Sahiwal, Red Sindhi, Gir and Deoni.** The milk production of milk breeds is on the average more than 1600 kg per lactation

b) Dual purpose breeds

The cows in these breeds are average milk producers and male animals are very useful for work. Their milk production per lactation ranged from 500 kg to 150 kg. The examples of this group are **Ongole, Hariana, Kankrej, Tharparkar, Krishna valley, Rathi, Gaolao and Mewati.**

c) Draught reeds

The male animals are good for work and Cows are poor milk yielders. Their milk yield as an average is less than 500 kg per lactation. They are usually white in colour. A pair of bullocks can haul 1000 kg. Net with an iron typed cart on a good road at walking speed of 5 to 7 km per hour and cover a distance of 30 - 40 km per day. Twice as much weight can be pulled on pneumatic rubber tube carts. The examples of this group are **Kangayam, Umblacherry, Amritmahal, Hallikar**. Draught breeds are again classified as follows:

1. **Short Horned:** White (or) Grey with long coffin shaped skull - Nagori, Bachur.
2. **Lyre Horned:** Grey with wide forehead - Malvi, Kherigarh.
3. **Small black, red** cattle with large patches of white marking found in Himalayan Region - Ponwar, Siri.
4. **Mysore type**: Prominent forehead with long and pointed horns which rise close together - Hallikar, Umbalachery, Alambadi, Pulikulam, Amritmahal, Burgur, Khillari, Kangayam.

Exotic cattle breeds

They offer a diverse range of characteristics and qualities, catering to specific agricultural needs and preferences. In this article, we will explore two main categories: **dairy breeds and beef breeds. Dairy breeds** are known for their exceptional milk production, e.g. Holstein Friesian, Jersey, Ayrshire, Guernsey, Red Dane, Brown Swiss, Dexter, Jamaica Hope, Dutch Belted etc.

While **beef breeds** are developed for efficient meat production. By delving into the distinctive traits, origins, and advantages of each breed, we aim to provide a comprehensive overview of these exotic cattle breeds. Few examples are Hereford, Short horn, Polled, short horn, Galloway, Aberdeen Angus, Brahman, Beef master, Santa Gertrudis etc.

Description of exotic dairy cattle breeds:

1. Jersey

Fig. 1: Jersey breed: bull and cow

- Originated from **Jersey Island, U.K.**
- One among the most popular dairy breeds.
- Smallest of the dairy types of cattle.
- In India this breed has acclimatized well and is widely used in cross breeding with indigenous cows.
- The **typical colour of Jersey cattle is reddish fawn**.
- Dished forehead; compact and angular body.
- **Economical producers of milk with 4.5% fat.**
- Average milk yield is 4500 kgs per lactation.

2. Holstein Friesian

Fig. 2: Holstein Friesian breed: bull and cow

- Originated from the northern parts of the Netherlands, especially in the province of Friesland and in Schleswig-Holstein in northern Germany.
- Holstein Friesian has the **highest milk production of all breeds worldwide**. The Holstein Friesian is the most widespread cattle breed in the world.
- Largest dairy breed and ruggedly built is shape and possesses large udders. Breeds have typical markings of black and white that make them easily distinguishable.
- The average **milk production of cows is 6000 to 7000 kgs per lactation**. The fat percentage in HF cow milk is only 3 to 3.5 percent.

3. Brown Swiss

- The mountainous region of Switzerland is the place of origin of the Brown Swiss breed.
- Breeds are rugged in nature and good milk production.
- Average milk yield is 5000 kgs per lactation.

Fig. 3: Brown Swiss breed: bull and cow

- Brown Swiss robustness allows it to **adapt to any climate: cold winters to warm climates.** Several scientific studies have demonstrated this ability. By evacuating the heat more easily and maintaining physical condition, **Brown Swiss is more resistant to heat than Holstein.**

4. Red Dane

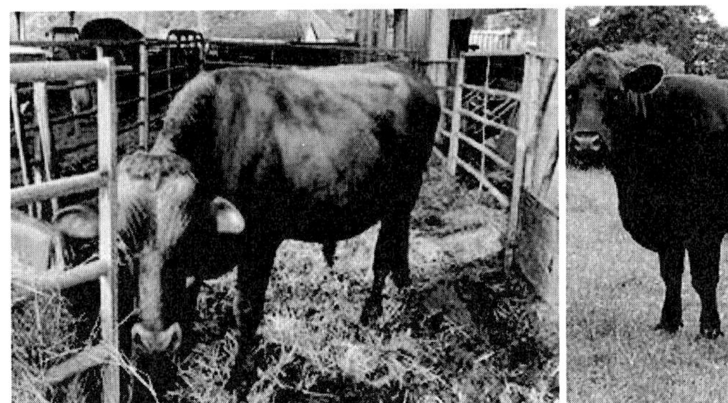

Fig. 4: Red Dane breed: bull and cow

- Originated in Denmark.
- Body colour of this Danish breed is red, reddish brown or even dark brown.
- It is also a heavy breed.
- The **lactation yield of Red Dane cattle varies from 3000 to 4000 kgs.**

5. Ayrshire

- Origin is Ayrshire in Scotland and is considered **the most beautiful dairy breed.** These are very active animals but hard to manage.

Fig. 5: Ayrshire breed of cow

- They do not produce as much milk or butter fat (only 4%) as some of the other dairy breeds.
- The breed was also known as **Dunlop cattle or Cunningham cattle.**

6. Guernsey

Fig. 6: Guernsey breed of cow

- Originated from the small Island of Guernsey in France. It has been suggested that the Guernsey derives from cattle imported from the French mainland, **brindled cattle from Normandy,** and wheaten stock similar to the **Froment du Léon of Brittany**. There may also have been some influence from Dutch cattle in the 18th century.
- The coat is red or fawn (wheat-coloured), and may or may not be multi-coloured red-and-white or fawn-and-white.
- The **milk has a golden colour due to an exceptionally high content of beta carotene**
- Guernsey cows produce around 6000 kgs per lactation.
- The Guernsey cow has many notable advantages for the dairy farmer over other breeds including high efficiency of milk production, low incidence of calving difficulty and longevity.

Exotic beef breeds

1. Angus

Originated in northeastern Scotland. Angus can be black or red and are polled cattle, or without horns. Oregon ranchers raise a large percentage of Black Angus cattle.

Fig. 7: Angus breed of bull

2. Hereford

Originated in England and became popular in the United States due to their early maturity and fattening abilities. Herefords live longer than many other breeds. They have a good temperament, are very fertile, and produce a lot of milk. Their coats are dark red and their faces are white.

Fig. 8: Herford breed of bull

3. Limousin

Originated in western France. Ranging in colour from red to gold, this breed can also be black.

Fig. 9: Limousin breed of cow

4. Shorthorn

Originated in northern England. This is a dual-purpose breed, meaning they are raised for both meat and milk. Their colour patterns range from red to white. They can be spotted or roan.

Fig. 10: Shorthorn breed of cow

5. Charolais

Originated in central France. This breed is known for its white to cream coloured coat.

Fig. 11: Charolais breed of cow

6. American Brahman

- Originated in India. These cattle can tolerate high heats, sunlight and are insect or parasite resistant because of their thick skin.
- The Brahman is an American breed of zebuine-taurine hybrid beef cattle. It was bred in the United States from 1885 using cattle originating

in India, imported at various times from the United Kingdom, India, and Brazil.

- Brahman, any of several varieties of cattle originating in India and crossbred in the United States with improved beef breeds, producing the hardy beef animal known as the American Brahman.
- Similar blending in Latin America resulted in the breed known as Indo-Brazil.
- Indian cattle were first imported into the Western Hemisphere in the mid-nineteenth century. The Gir, Gujarat, Nellore, Krishna Valley and Ongole breeds were particularly successful in the southern United States and in Brazil, where heat, humidity, and pests made northern European breeds less profitable.
- The Hereford and the Shorthorn were among the first breeds used in crossing and have remained popular. Beef of these mixtures, such as the Beefmaster, is markedly low in fat.
- Other prominent crosses include the Charbray, from the Brahman and Charolais, and the Brangus, from the Brahman and Angus.
- Pure-bred Brahmans today are used mostly for breeding and seldom slaughtered.

Fig. 12: American Braham breed: bull and cow

Description of Indian cattle breeds:

1. Sahiwal

Fig. 13: Sahiwal breed: bull and cow

- **Best indigenous dairy breed.**
  - Originated in the Montgomery region of undivided India, it now belongs to Pakistan.
  - This breed is otherwise known as **Lola (loose skin), Lambi Bar, Montgomery, Multani, Teli.**
  - The breeding tract of the breed is Ferozpur and Amritsar districts of Punjab and Sri Ganganagar district of Rajasthan.
  - The cows are brownish red in colour; shades may vary from a mahogany red brown to more greyish red.
  - Heavy breed with a symmetrical body having loose skin.
  - The average milk yield of this breed is between 2325 kg, ranges from 1600 and 2750 kg per lactation with 4.9% milk fat content (range 4.8 to 5.1%).
  - Because of its desirable traits, it is being utilised widely in many warm humid countries of the world for improvement of local stock or for initial crossbreeding of indigenous stock before undertaking upgrading with European breeds.
  - This breed is widely used for grading the non-descriptive cattle population of India in different states.
  - Milk yield as high as 6000 litres has been recorded under organized farm conditions. Sahiwal cattle milk has a high-fat content of 5-6%. Considering the merit of this breed, Sahiwal animals were imported

by Australia and were used in developing a synthetic crossbred called Australian Milking Zebu (AMZ) cattle.

- Remarkably, the record for the highest milk yield is held by a Sahiwal cow named 'MUDINI' from the Military Farms Department of the Indian Army. She produced over 6000 kg of milk in 440 days during an extended lactation.
- The population size of this breed 59,49,674 as per 20th Livestock census, 2019. It is about 4.2% among all breeds of India or rank third in total number of breeds.

2. Gir

Fig. 14: Gir breed: bull and cow

- The breed is also known as "Bhodali", "Desan", "Gujarati", "Kathiawari", "Sorthi", and "Surati".
- Originated in **Gir forests of South Kathiawar in Gujarat also found in Maharashtra and adjacent Rajasthan**.
- Basic colours of skin are white with dark red or chocolate-brown patches or sometimes black or purely red.
- **Horns are peculiarly curved, giving a 'half-moon' appearance**.
- Horns are peculiarly curved. Starting at the base of the crown they take a sideways downward and backward curve and again incline a little upward and forward taking a spiral inward sweep, ending in a fine taper, thus giving a half-moon appearance.
- Long and pendulous ears are folded like a leaf. Ears hang all the time and their inside face forward.

- Animals are maintained in semi-intensive management systems, which are largely bred by professional breeders known as Rabaris, Bhanoads, Maldharis, Ahirs and Charans.
- Average lactation yield of Gir cows is 2110 kg with an average milk fat of 4.6 % (ranges from 3.9 to 5.1 %).
- Milk yield ranges from 1200-1800 kgs per lactation.
- This bread is known for its hardiness and disease resistance.
- A record production of 3182 kg of milk at 4.5% fat has been achieved by Gir cows in India.
- The population size of this breed 68,57,784 as per 20th Livestock census, 2019. It is about 4.8% among all breeds of India or rank first in total number of breeds.

3. Red Sindhi

Fig. 15: Red Sindhi breed: bull and cow

- This breed is otherwise called Red Karachi and Sindhi and Malir.
- Originated in **Karachi and Hyderabad (Pakistan) regions of undivided India** and also reared in certain organized farms in our country.
- It is considered that the breed evolved from Las Bela cattle of Bela, Balochistan.
- Colour is red with shades varying from dark red to light, strips of white.
- Bullocks, despite being lethargic and slow, can be used for road and field work.
- Occasionally small white patches are seen in dewlap and forehead.
- Horns are thick at the base and emerge laterally and curve upward.

- The Red Sindhi breed has a very high genetic potential for milk production and is comparable with Sahiwal.
- In India, the animals of the breed are not available in field conditions. The breed was used in many countries including the USA, Australia, Philippines, Brazil and Sri Lanka for breed development.
- The milk yield of the cattle ranges from 1100 to 2600 kg per lactation with an average yield of 1840 kg per lactation.
- Fat percentage in the milk varies from 4 to 5.2% with an average of 4.5%.
- In well-managed herds, Red Sindhi cows can yield up to 5,450 kg of milk in a lactation period of a little over 300 days.
- The population size of this breed 6,12,900 as per 20th Livestock census, 2019. It is about 0.4% among all breeds of India.

4. Tharparkar

- Tharparkar (named after the Thar Desert in Rajasthan) is a dual purpose and disease resistant cattle breed.

Fig. 16: Tharparkar breed: bull and cow

- Originated in **Tharparkar district (Pakistan) of undivided India and also found in Rajasthan.**
- Otherwise known as **White Sindhi, Gray Sindhi and Thari**.
- The breeding tract of the breed includes Kachchh district of Gujarat and Barmer, Jaisalmer and Jodhpur districts of Rajasthan.
- They are medium-sized, compact and have lyre-shaped horns.

- Body colour is white or light grey.
- The bullocks are quite suitable for ploughing and casting and the cows yield 1800 to 2600 kgs of milk per lactation.
- Tharparkar cows are excellent milk producers. On average, you can expect them to yield 10 to 14 Liters of milk per day.
- Average milk fat percentage is 4.88% which ranges from 4.72 to 4.90%.
- Good animals with high nutrition have produced even higher than 3000 litre per lactation in farm conditions. The males are also good for drought purposes. Due to better heat tolerance and disease resistance, this breed was used for producing "Karan Fries", a synthetic crossbred cattle breed at National Dairy Research institute (NDRI), Karnal, Haryana.
- Tharparkar cows typically have a lifespan of around 25 years and give birth to 14-18 calves during their lifetime.
- The population size of this breed 5,82,257 as per 20th Livestock census, 2019. It is about 0.4 % among all breeds of India.

5. Hariana

- Hariana is one of the most prominent dual-purpose cattle breeds of the Indo Gangetic plain and named according to the breeding tract of the breed (Haryana state). The breed was earlier known as 'Hisar' and 'Hansi' according to their place of origin.
- The breeding tract of the breed includes Hisar, Rohtak, Sonipat, Gurgaon, Jind and Jhajjar districts of Haryana. This breed also popular in Punjab, Uttar Pradesh and parts of Madhya Pradesh.

Fig. 17: Hariana breed: bull and cow

- Typically, the breed is white or light grey coloured with coffin shaped skulls. In bulls colour in between fore and hind quarters is relatively dark or dark grey.
- The animals have long and narrow faces, well-marked bony prominence at the centre of poll and small horns.
- The breed is mainly maintained for bullock production as they are powerful work animals and therefore more attention is paid in managing male calves.
- However, the cows also produce a fair amount of milk. Good cows can produce even up to 1700 kg of milk in a lactation with average cows producing around 997 kg in a lactation (ranging between 693 to 1745 kg).
- The milk yield of Hariana cows varies from 2000 kg to 4000 kg per lactation, with fat content varying from 4% to 4.5%.
- Scientists from Lala Lajpat Rai University of Veterinary and Animal Sciences (LUVAS), Hisar, Haryana managed to obtain a record 20.6 kg of milk in a single day from a Hariana cow.
- The population size of this breed 27,57,186 as per 20th Livestock census, 2019. It is about 1.9 % among all breeds of India.

6. Kankrej

Fig. 18: Kankrej breed: bull and cow

- Kankrej cattle are also known as "Wadad" or "Waged", "Vagadia", "Talabda", "Nagar", "Bonnai". It takes its name from the name of geographical area i.e. Kankrej taluka of Banaskantha district in Gujarat.
- They are found in the area southeast of Rann of Kutch comprising Mehsana, Kachchh, Ahmedabad, Kheda, Anand, Sabarakantha and Banaskantha districts of Gujarat and Barmer and Jodhpur districts of Rajasthan.

- Originated from **Southeast Rann of Kutch of Gujarat and adjoining Rajasthan** (Barmer and Jodhpur district).
- **The horns are lyre-shaped.**
- The colour of the animal varies from silver-grey to iron-grey or steel black.
- **The gait of Kankrej is peculiarly called as 1 ¼ paces (Sawai chal).**
- The unique characteristics like resistance to tick fever, heat stress, very little incidence of contagious abortion and tuberculosis made Kankrej a very popular one among these other breeds.
- Kankrej is valued for fast, powerful, draught cattle. Useful in **ploughing and carting**.
- Coat colour of the animal varies from silver grey to iron grey and steel black. In males, the forequarters, hindquarters, and hump are slightly darker than the rest of the body.
- Bulls tend to get darker than cows and bullocks. The hump in the males is well developed and not as firm as in other breeds.
- Forehead is broad and slightly dished in the centre. Face is short and nose slightly upturned.
- Unique characteristic of this breed is its large, pendulous ears. **The horns are lyre shaped very big in size and twisted which look very attractive and distinct from other breeds.**
- The cows are good milkers. The cows yield on an average 1738 Kg and a maximum of 1800 Kg of milk in a lactation. Bullocks are used for agricultural operations and road transport.
- Best cows produced around 4900 kg milk in a lactation at village conditions.
- Milk Fat percentage: Ranged from 4.7-5% with average of 4.8% (NBAGR).
- Kankrej, famously known as Guzerat in Brazil, is being maintained in large numbers as a pure breed in that country.
- The population size of this breed 22,15,537 as per 20th Livestock census, 2019. Its proportion is about 1.6 % among all breeds of India.

7. Rathi

- **Rathi**, also synonymous with its variant Rath, is a breed of cattle indigenous to India. It originated in the region of the state of Rajasthan consisting of Bikaner, Ganganagar and Hanumangarh districts.

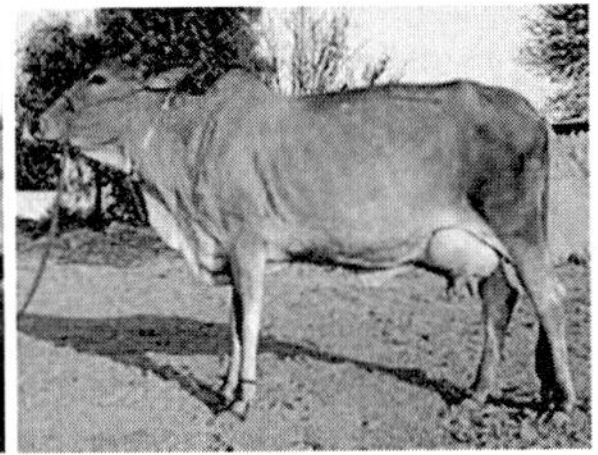

Fig. 19: Rathi breed: bull and cow

- It is an important dual-purpose cattle breed of India known for both its milking and draught prowess. The cattle are locally known to have two variants, **Rathi is a draught breed, while Rath is a pure milch variant**.
- The Rath variant originated in the Alwar district of Rajasthan and was domesticated by **Rath tribe**.
- Rath is characterized by its white skin with black or grey spots while Rathi is **usually of brown colour.**
- The Rathi cattle breed is used for dual purposes. The cows of this breed are one of the higher milk producing breeds and are efficient in terms of producing higher milk yield. The cows of this breed reared for producing milk and males utilized for draught purposes.
- When it comes to milk production capacity, Rathi cows have proved their worth. In some cases, they have given tough competition to Sahiwal cows as well.
- On average, Rathi cows yield 5-10 litre milk per day. There were some rare instances when Rathi cows produced 15-20 litre milk per day. The lactation milk yield of this cattle breed ranges from 1500-3000 litre.
- Selected cows have produced around 4800 kg per lactation at farmer's doorstep.
- The fat content in Rathi cow milk ranged from 4.4% to 9%.
- In general, milk from *Bos indicus* cattle (including Rathi) can have fat content as high as 5.5%.

- The population size of this breed 11,69,828 as per 20th Livestock census, 2019. Its share about 0.8 % among all breeds of India.

Special features of Rathi cattle breed

- These creatures are strong enough to withstand injury and illness and survive amidst unfavourable circumstances.
- They travel many kilometres in a day as they are strong from the inside.
- The temperature of Rajasthan easily touches the mark of 50°C. Despite the scorching heat, they live and survive with ease.
- The ecosystem in its natural habitat is considered fragile.
- These cows stay in regions known for dry monsoons, cold winters, hot and sweltering summers, and dust storms. The Rathi cattle breed has embraced this habitat and produces milk efficiently against all the odds.
- Rathi cattle are thought to have evolved from **intermixing of Sahiwal, Red Sindhi, Tharparkar and Dhanni breeds** with a preponderance of Sahiwal blood.
- They are **known as the Queen of Rajasthan.**

8. Ongole

Fig. 20: Ongole breed: bull and cow

- Otherwise known as **Nellore,** as the Ongole area was earlier in **Nellore district.**
- Ongole cattle are an important indigenous cattle breed that originates from **Prakasam District** in the state of Andhra Pradesh in India.
- The breed derives its name from the place the breed originates from, Ongole.

- The breeding tract of the breed includes **East Godavari, Guntur, Ongole, Nellore and Kurnool districts of Andhra Pradesh** and extends all along the coast from **Nellore to Vizianagram.**
- The Ongole breed of cattle is in great demand as it is said to possess **resistance to both foot and mouth disease and mad cow disease.**
- These cattle are commonly used in bull fights in Mexico and some parts of East Africa due to their strength and aggressiveness.
- The Ongole was transported to several western countries including Brazil and USA.
- Animals of the Ongole breed were extensively exported to the USA for beef production; Brazil for beef and milk production; Sri Lanka, Fiji and Jamaica for draught; Australia for heat tolerance and beef; and Switzerland for disease resistance.
- They also participate in traditional bull fights in Andhra Pradesh and Tamil Nadu.
- Cattle breeders use the fighting ability of the bulls to choose the right stock for breeding in terms of purity and strength.
- The mascot of the 2002 National Games of India was Veera, an Ongole bull.
- Home tract is **Ongole taluk in Guntur district of Andhra Pradesh**.
- Large muscular breed with a well-developed hump.
- The breed has a glossy white coat colour. Males possess dark markings on head, neck and hump and black points on knees and pastern.
- Horns are short and stumpy, growing outward and backward from the outer angles of the poll, thick at the base and firm without cracks.
- In cows, horns are thinner than bulls. Horns in cows generally extend outward, upward and inward. Breed can be identified visibly by its majestic gait, stumpy horns and large fan shaped and fleshy dewlap.
- Suitable for heavy draught work.
- White or light grey in colour.
- Average milk yield is 1000 kg per lactation and with an average fat percentage of 3.79%.
- The population size of this breed 7,03,142 as per 20th Livestock census, 2019. It's share about 0.5 % among all breeds of India.

9. Krishna Valley

Fig. 21: Krishna Valley breed: bull and cow

- Originated from black cotton soil of the watershed of the river Krishna in Karnataka and also found in border districts of Maharashtra.
- The breeding tract of the breed includes Belgaum, Raichur and Bijapur districts of Karnataka and Satara, Sangli and Solapur districts of Maharashtra.
- **It is believed that the Gir and Kankrej breeds of Gujarat, Ongole breed of Andhra Pradesh and local cattle having Mysore type blood** in them have contributed to the origin of this breed.
- Its massive body and distinct bulging forehead can easily identify this breed. Horns are small in size and curved and usually emerge in an outward direction from the outer angles of the poll curving slightly upward and inward.
- Animals are large, having a massive frame with a deep, loosely built short body.
- Tail almost reaches the ground.
- Generally grey white in colour with a darker shade on fore quarters and hind quarters in males. Adult's females are more whitish in appearance. Brown and white, black and white, and mottled colours are often seen.
- The bullocks are powerful animals useful for slow ploughing, and valued for their good working qualities.
- The average yield is about 900 kg per lactation.
- In Karnataka, the average milk fat percentage of these cattle is reported to be 5.7%, but individual animals within a breed can still exhibit variations.
- The population size of this breed 22,532 as per 20th Livestock census, 2019.

10. Deoni

Fig. 22: Deoni breed: bull and cow

- This breed is otherwise known as Dongerpati, Dongari, Wannera, Waghyd, Balankya, Shevera.
- Originated in **Western Andhra Pradesh and also found in Marathwada region of Maharashtra state and adjoining part of Karnataka**.
- It has evolved from a strain descended from a **mixture of Gir, Dangi and local cattle.** The breed derived its name from **Deoni taluka of Latur district of Maharashtra**.
- The breeding tract lies in the Balaghat range of Sahyadri hills extending from Kannad taluka of Aurangabad to Deglur taluka of Marathwada region of Maharashtra state. The actual place of origin is Deoni, Udgir and Ahmadpur taluka of Latur district.
- Body colour is usually spotted black and white.
- Body colour is usually spotted black and white. This breed has three strains viz. Balankya (complete white), Wannera (complete white with partial black face) and Waghyd or Shevera (black and white spotted).
- Bullocks are suitable for large cultivation.
- Small sized horns emerge from the side of the poll behind and above the eyes in outward and upward direction. The tips of the horn are blunt.
- The breed is characterized by drooping ears and prominent and slightly bulging forehead. Deoni bullocks are preferred for heavy works and bullocks can effectively be used even up to 12 years of age.
- Milk yield ranged from 636 to 1230 kg per lactation with an average milk fat is 4.3% (ranges from 2.5 to 5.3 %).

- Caving interval averaged 447 days.
- The population size of this breed 2,84,342 as per 20th Livestock census, 2019. It's share about 0.2 % among all breeds of India.

11. Hallikar

Fig. 23: Hallikar breed: bull and cow

- Also known as "**Mysore**", the breed is considered as **best draught breed of Southern India.**
- Originated from the former princely state of Vijayanagarm, presently part of Karnataka.
- The breeding tract comprises Mysore, Mandya, Bangalore, Kolar, Tumkur, Hassan and Chitradurga districts of Karnataka.
- The colour is grey or dark grey.
- Compact, muscular and medium size animal with prominent forehead, **long horns and strong legs.**
- The breed is best known for its draught capacity and especially for its trotting ability.
- Young breeding bulls have dark shades on the shoulder and hindquarters.
- **Horns emerge near each other from the top of poll and** are carried straight, upward and backward, till nearly half the length and then orient slightly forward and inward with pointed tips.
- White markings or irregular patches around the eyes, cheeks, neck or shoulder region are also found.

- The animals are maintained in a semi-intensive management system by professional breeders. Green fodder chiefly comprises finger-millet, grass, sorghum or pearl-millet.
- Average milk yield per lactation is 542 kg ranged from 227-1134 kg with average milk fat of 5.7%.
- The population size of this breed 8,19,137 as per 20th Livestock census, 2019. It's share about 0.6 % among all breeds of India.

12. Amritmahal

Fig. 24: Amritmahal breed: bull and cow

- Originated in **Hassan, Chikmagalur and Chitradurga districts of Karnataka**.
- Amritmahal animals are grey but their body colours vary from almost white to near black.
- The muzzle, feat and tail are usually black.
- **Horns are long and end in sharp black points.**
- Amritmahal is also known as "Doddadana", "Jawari Dana" and "Number Dana". "Amrit" means milk and "Mahal" means house.
- The breeding tract includes Chikmagalur, Chitradurga, Hassan, Shimoga, Tumkur and Davanagere districts of Karnataka.
- Amritmahal was developed from a herd established by the ruler of Mysore state between 1572 and 1636 A D and was **developed from a draught breed of southern India with an objective to increase the milk productivity.**
- The breed is usually Grey in colour but the colour varies from white to almost black. White grey markings are present on the face and dewlap in some animals. Dark shades on neck, shoulder, hump and hindquarters.

- Head is long and tapers towards the muzzle.
- Horns are long and emerge from the top of the poll, fairly close together in backward and upward direction, turn in and end in sharp black points, sometimes touching each other.
- This is a famous draught breed known for its power and endurance and animals are fiery and active.
- **Bullocks are especially suited for trotting and fast transportation.**
- Cows are very poor milkers. Average milk yield per lactation is 572 kg.
- Milk Fat Content: The milk contains an average fat content of 5.7%.
- The population size of this breed 3,01,354 as per 20th Livestock census, 2019. It's share about 0.2 % among all breeds of India.

13. Khillar

Fig. 25: Khillar breed: bull and cow

- Originated from **Sholapur and Sitapur districts of Maharashtra.**
- Closely resembles **Hallikar breed.**
- Synonyms: Mandeshi, Shikari, Thillar
- Grey-white in colour.
- **Long horns turn forwards in a peculiar fashion**. The horns are generally black, sometimes pinkish.
- Bullocks are fast and powerful.
- The breeding tract of this breed is Belgaum, Bijapur, Dharwad, Gulbarga,

Bangalkote districts of Karnataka and Pune, Satara, Sangli, Solapur, Kolhapur, Osmanabad districts of Maharashtra.

- **This breed is believed to have originated from the Hallikar or Amritmahal breed of cattle.**
- Khillar means herd of cattle and herdsman is known as Khillari or Thillari, in the Satpura range of hills, he is known as Thillari. There is a special tribe of professional cattle breeders in this region known as Thillaris.
- **Four types of Khillar cattle are prevalent** in different parts of the country. "Atpadi Mahal" in Southern Maharashtra, "Mhaswad" in Solapur and Satara areas, "Thillari" in Satpura range of hills, and "Nakali" in adjoining areas of this region.
- The typical Khillar animal is compact and tight skinned, with clean cut features. The whole appearance is like a compact cylinder with stout, strongly set limbs. There is a slight rise in the level of the back towards the pelvis.
- Horns are long and pointed and follow the backward curve of the forehead which are placed close together at the root with thick base, grow backward for half of the length and then turn upwards in a smooth bow shape peculiar to this breed ending in pointing tips.
- The ribs are well sprung and give the trunk a barrel shape. The gait of Khillar is quick and spirited. Khillar cattle of Deccan plateau, the "Mhaswad" and the "Atpadi Mahal" type are greyish-white.
- Males are dark over the fore and hindquarters with peculiar grey and white mottled markings on face. The "Tapti Khillar" are white with carroty nose and carroty hooves.
- Bullocks are highly valued as fast powerful draught animals. They can travel miles without showing any signs of fatigue.
- Average milk yield per lactation of Khillar cattle is 451 kg with an average milk fat of 4.22 %. The lactation yield ranged from 240 to 515 kg.
- Cattle of this breed have been exported to North-Western Sri Lanka to improve the draught qualities of Sinhala breed.
- The population size of this breed 12,99,196 as per 20th Livestock census, 2019. It's share about 0.9 % among all breeds of India.

14. Kangayam

Fig. 26: Kangayam breed: bull and cow

- Also known as **Kongu and Konganad.**
- The breeding tract of this breed is **Coimbatore, Erode, Dindigul, Karur and Namakkal districts of Tamil Nadu**. This breed derives its name from its habitat, **Kangayam taluk of Erode district** (earlier part of Coimbatore district).
- Bulls are grey with dark colour in hump, fore and hind quarters.
- Cows are **grey or white.**
- The eyes are dark and prominent with black rings around them.
- Moderate size with compact bodies.
- This breed resembles the Umblachery breed. Animals true to the type are available in Kangayam and Dharapuram taluks of Erode district, and Karur taluk of Karur districts.
- The breed is usually of **grey or white in colour**. Coat is red at birth, but changed to grey at about six months of age.
- The males are generally grey in colour with black or **very dark grey markings on the head, neck, hump and quarters.**
- Cows are generally white and grey with deep markings on the knees, and just above the fetlocks on all four legs. The most prominent feature of this breed is that they **have dark eyes with prominent black rings around them.**

- Horns are long and strong, take backward, outward and upward sweep and then curving **inward with tips tending to meet each other to form crescent shape.**
- Bullocks have good capacity for work and are used for various agricultural operations and for works like sugarcane load hauling.
- Average lactation milk yield of Kangayam cattle is 540 kg with an average milk fat of 3.9% (ranged from 1.6 to 7.7 %).

15. Bargur

Fig. 27: Bargur breed: bull and cow

- Found around **Bargur hills in Bhavani taluk of Erode district in Tamil Nadu.**
- Developed for work in uneven hilly terrains.
- **Generally brown colour with white markings.**
- Animals are well built, compact and medium in size.
- Known for their speed and endurance in trotting.
- Cautious in behaviour and tends to remain away from strangers.
- **Horns are light brown in colour and emerge closer at the root and are inclined backward, outward and upward with a forward curve, which is sharp at the tip.**
- The animals are maintained in an extensive management system and animals are reared in forest areas in semi-wild conditions and housed in enclosures called "Pattys" in groups of 50 to 200 heads by hired local tribal labourers called "Lingaiys".

- The average milk yield per lactation is reported as 350 kg, which ranged from 250-1300 kg.
- Fat percentage in Bargur cows milk ranged from 4.7 to 5.3%.

16. Umblachery

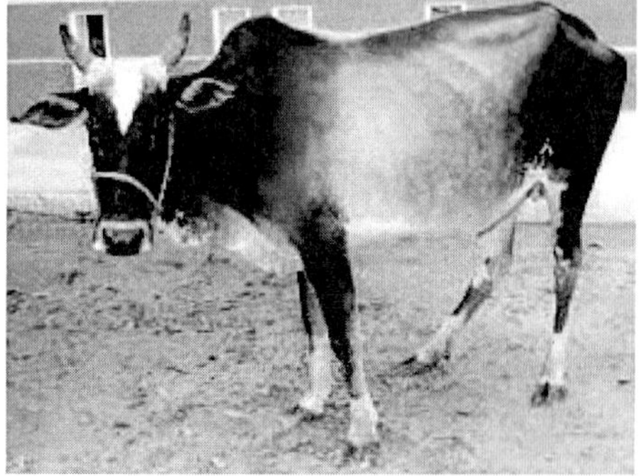

Fig. 28: Umblachery breed: bull and cow

- Originated in Thanjavur, Thiruvarur and Nagappattinam districts of Tamil Nadu.
- Suitable for wet ploughing and known for their strength and sturdiness.
- **Umblachery calves are generally red or brown at birth with all the characteristic white markings on the face, limbs and tail.**
- The practice of dehorning of bullocks is peculiar in Umblachery cattle.
- A noted draught cattle breed of Tamil Nadu, Umblachery is famous for its sturdiness and strength.
- It derives its name from a small village with the same name (Umblachery) considered as native tract of the breed. It is also known as "Jathimadu", "Mottaimadhu", "Molaimadhu", "Southern", "Tanjore" and "Therkuthimadhu".
- Typically, the animals are red at birth and thereafter changed to grey during development.
- Males are dark grey with black extremities, whereas, cows are grey with light dark grey area in face, neck and hip regions.
- There are white markings on face, limb and tail and the marking on the leg resembles socks.

- This breed is suitable for ploughing, carting, threshing and levelling in marshy paddy fields because of its medium size.
- Most of the cows are not milked and calves are allowed to suckle the dam. Average recorded milk yield is 494 kg per lactation with an average milk fat of 4.94%.

17. Pulikulam

Fig. 29: Pulikulam breed: bull and cow

- This breed is commonly seen in Cumbum valley of Madurai district in Tamil Nadu.
- Also known as **Jallikattu madu, kidai madu, sentharai**.
- Small in size, usually grey or dark grey in colour.
- Mainly used for penning in the field.
- Presence of reddish or brownish spots in muzzle, eyes, switch and back is the characteristic feature of this breed.
- Typical backward curving horns like Mysore type cattle.
- The animals of this breed are active, useful for draught and ploughing but not fast trotters. Pulikulam is a **popular draught and game breed** of Tamil Nadu. It is named after its village of origin (Pulikulam in Tamil Nadu) and is also known as "Palingu maadu", "Mani maadu", "Jallikattu maadu", "Mattu maadu" and "Kilakattu maadu".
- Madurai, Sivaganda and Virudhunagar districts of Tamil Nadu form its breeding tract.
- Males are dark grey in colour while females are white or grey.

- The animals are small sized with a compact body and short legs. Muzzle, eyelids, tail switch and hooves are black. Forehead is broad and has a groove at the centre.
- Hump is large in males and small in females. Udder is not well developed.
- Horns are curved outward, upward, backward and then inward, ending with pointed tips spaced wide apart.
- The breed is not reared for milk production and only a few animals are milked in a large herd for consumption by herdsmen.
- Generally, milk yield is about 1.25 kg per day, which varies from 0.5 kg to 2 kg. The animals are also reared for manure purposes.
- The population size of this breed is 13,934 as per 20th Livestock census, 2019.

18. Alambadi

Fig. 30: Alambadi breed: bull and cow

- Alambadi is one of the five indigenous cattle breeds of Tamil Nadu. This breed is generally present on the bank of Cauvery River. Population of this breed is dwindled considerably in its native tracts in Dharmapuri, Krishnagiri and Salem district of Tamil Nadu namely Alambadi, Pikkili, Periyur, Hamanur, Kottur, Pannapatti, Pucharampatti, Jambut, Pudhinatham, Urigam, Anchetty, Doddamanchi, Bettamugilalam and Natrampalayam villages.
- Originated from **Alambadi of Dharmapuri district in Tamil Nadu.**
- White markings will be seen on the forehead, limb and tail and sometimes on legs.

- Horns are backward curving like Mysore cattle.
- Resembles Hallikar and is also known as Betas.
- It is useful in ploughing.
- This hardy breed is considered a medium-sized animal and has a greyish-white coat with reddish-brown heads and backs.
- Its ears are pendulous longer than other breeds which is known to have some heat-dissipating benefits.
- The average weight of an Alambadi bull is 350 kg while a heifer weighed up to 300 kg.
- Historically this breed has been able to survive in areas with scant rainfall due to mental hardiness as well as resilience against diseases without much interference from humans.
- Historically, they were kept as draught animals, but modern breeders are attempting to increase their value through their breeding practices. Advanced technologies such as genomics and artificial insemination have been employed by some ranchers to produce animals with desirable traits.
- The Alambadi breed of cattle are quite rare and on the endangered species list in India.
- The Alambadi bull has a much larger frame than the cow does and is rippled with more muscle.
- They both had a long slender head that resembles a long heart shape.
- They have a tight dewlap and a small cervices-thoracic hump on their shoulders.
- Cows only calve once a year and should have a 12 to 14 months inter-calving cycle. The Alambadi cow is what they term a very bad milker which means that she is not used for her milk. They are known to produce large quantities of milk, but the quality of her milk is not good at all. They are good mothers and will wean their young.
- Milk yield per animal is limited to 1-3 litres per day but they are known for their draught ability.
- Their milk is characterized by its **rich quality and high-fat content**, and the milk is highly regarded for producing exceptional dairy products like ghee.

- Alambadi cattle breed was first documented by W.D. Gunn during 1909 and further documentation done by R.W. Littlewood during 1936. This breed derived its name, Probably, from the village called Alambadi on the bank of river Cauvery. This breed is also called as "Beslal or Cauvery breed", "Salem", "Mahedeswarabetta", "Cauvery valley cattle ".
- In order to conserve as well as to undertake breed improvement programmes of Alambadi Cattle, Government of Tamil Nadu has sanctioned Rs 400 lakhs to establish Alambadi Cattle Breed Research Centre at Dharmapuri district during the year 2018-2019.

19. Bachaur

- Bachaur is a draught breed of Bihar which is also known as "Bhutia".
- The breeding tracts of Bachaur cattle are Sitamarhi, Darbhanga and Madhubani districts of Bihar. Animals are now mostly concentrated in the areas adjoining Nepal border of India, comprising Bachaur and Koilpur sub-divisions of Sitamarhi district of Bihar State.

Fig. 31: Bachaur breed: bull and cow

- Common colours of Bachaur cattle are grey or greyish white.
- They are compact with straight backs. The forehead is broad and flat or slightly convex.
- The horns are medium-sized and stumpy, curving outward, upward and then downward. Bullocks can work for long periods without any break.
- Bachaur animals are mainly reared by Koir and Ahir communities in the breeding tract.
- Average milk yield per lactation of Bachaur cows is 347 kg (ranges from 225 to 630 kg) with an average milk fat of 5% (ranges from 4.8 to 7.1%).

20. Badri

- The Badri breed derived its name from the holy shrine of Char Dham at Badrinath.
- It is found only in the **hill districts of Uttarakhand** and was earlier known as the '**Pahadi**' cow.
- The breeding tracts include Nainital, Almora, Bageshwar, Pithoragarh, Champawat, Pauri Garhwal, Tehri Garhwal, Rudraprayag, Uttarkashi and Chamoli districts of Uttarakhand.
- Animals of this breed are well adapted to hilly terrains and climatic conditions and have balanced gait while walking.
- This cattle breed is small in size with long legs and varied body colours are black, brown, red, white or grey.
- Horns are curved-upward and inward. Badri is Small sized, active and sure-footed animal.
- Straight forehead with prominent poll, medium to large hump.
- Udder small and tucked up with the body. Badri is reared mainly for bullock power, milk and manure.
- Its urine has a high value due to its feeding and habitat. The lactation milk yield ranged from 547 to 657 kg, with an average milk fat content of 4% (Uttarakhand Livestock Development Board, 2022).

Fig. 32: Badri breed: bull and cow

21. Belahi

Fig. 33. Belahi breed: bull and cow

- Belahi breed of cattle are also known as Morni/ Desi.
- The breeding tracts of this breed lie in the **foothills of Shivalik in Haryana State and includes Ambala, Panchkula, Yamuna Nagar districts of Haryana and Chandigarh**.
- Term 'Belaha' is used to describe a mixture of colours. Animals of this breed are migratory in nature and are maintained on low input cost.
- Most common colours of the animals are reddish brown, Grey or white.
- Horns are sickle shaped, curving upwards and inwards.
- Animals are medium sized and have uniform but distinct body colour patterns.
- The face and extremities are white in colour and different degrees of white colour can be seen on the ventral part of the body.
- Head is straight and broad with prominent poll.
- The average standard lactation yield is 1014 kg ranging from 182 to 2092 kg with average milk fat of 5.25% ranging from 2.37 to 7.89%.
- The population size of this breed is 5,264 as per 20th Livestock census, 2019.

22. Binjharpuri

Fig. 34: Binjharpuri breed: bull and cow

- Binjharpuri, also known as "Deshi", is a breed of cattle found in **Jajpur, Kendrapara and Bhadrak districts of Odisha, maintained for milk, draught and manure.**
- Animals of this breed are medium sized, strong and chiefly white in colour. Few animals are grey, black or brown.
- In males, the hump, neck and some regions of face and back are black in colour irrespective of their coat colour.
- Horns are medium in size and curved upward and inward.
- The milk yield per lactation ranged from 915-1350 kg with milk fat ranging from 4.3-4.4%.

23. Dangi

Fig. 35.: Dangi breed: bull and cow

- The breeding tract of Dangi breed includes the **Dangs district of Gujarat and Thane, Nasik, Ahmednagar districts of Maharashtra.**
- The Dangi breed is also known as "**Kandadi**". Dangi cattle are extensively used for ploughing, harrowing and other field operations, and for carting timber from forest areas.
- The breed is well known for its excellent working qualities in heavy rainfall areas, rice fields and hilly tracts.
- The animals are well adapted to heavy rainfall conditions.
- **The skin exudes an oily secretion, which protects them from heavy rain.**
- Dangi cattle have a distinct white coat colour with **red or black spots** distributed unevenly over the body.
- Horns are **short and thick with lateral pointing tips**. Animals with inward pointing horns or downward pointing horn tips are also available in sizable numbers.
- Average milk yield per lactation is 430 kg with an average milk fat of 4.3%. The lactation milk yield ranged from 175 to 800 kg.

24. Gangatiri

Fig. 36. Gangatiri breed: bull and cow

- Gangatiri is also known as Eastern Hariana or Shahabadi.
- The breeding tracts include **Bhojpur district of Bihar and Varanasi, Mirzapur, Ghazipur and Ballia districts of Uttar Pradesh.**
- The animals of this breed are medium milk producers and possess

good draft ability also. The colour is complete white (Dhawar) or Grey (Sokan).

- The horns are medium-sized and emerge from the side of the poll behind and above eyes in outward and curving upwards and inwards ending with pointed tips.
- The forehead is prominent, straight and broad with a shallow groove in the middle. Eyelids, muzzle, hooves and tail switch are generally black in colour.
- The average milk yield in a lactation is around 1050 kg, varying from 900 to 1200 kg with an average fat of 4.9%, ranging from 4.1 to 5.2%.

25. Gaolao

Fig. 37: Gaolao breed: bull and cow

- Gaolao is a breed of western and middle part of India and known for its agility.
- The breed is also known as "Arvi" and "Gaulgani". The breeding tracts of the breed include Balaghat, Chhindwara, Seoni districts of **Madhya Pradesh**; Durg and Rajnandgaon districts of **Chattisgarh and Wardha and Nagpur districts of Maharashtra.**
- The breed is suitable for transportation in hilly areas.
- The breed resembles the Ongole breed, except that it is lighter.
- The coat colour is blackish white in males and white in females.
- Males are generally grey over the neck. Horns are short, stumpy and curved slightly backward.

- Head is markedly long and tapers towards the muzzle. Forehead recedes at the top giving a slightly convex appearance. Bullocks are trained for moving fast.
- The milk yield is low with an average of 604 kg per lactation with 4.32% fat (range between 470 to 725 kg per lactation).

26. Ghumusari

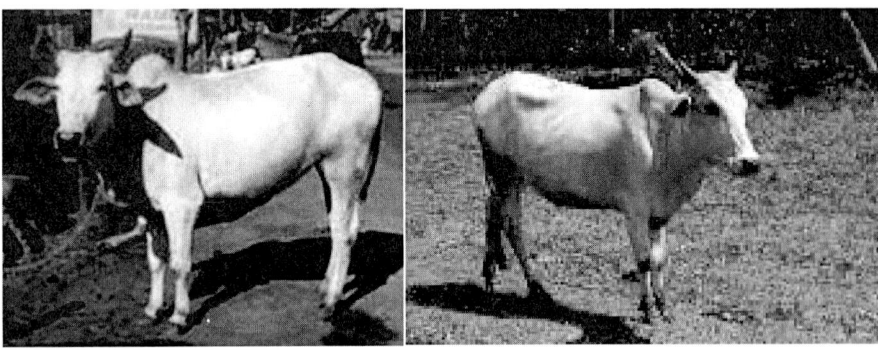

Fig. 38: Ghumusari breed: bull and cow

- Ghumusari breed of cattle is also known as "Deshi". Breeding tracts include western part of **Ganjam district and adjoining areas of Phulbani district of Odisha**.
- It is chiefly a draught cattle breed, but few animals are maintained for milk, manure and fuel. The draught ability of bullocks is considered superior to other breeds of bullocks in the native tract.
- The animals are small sized and strong in built. They are **chiefly white in colour**, but sometimes shades of grey are also visible.
- The horns are medium-sized, mostly curved upward and inward, but some animals have straight horns as well.
- The animals of this breed have a small head with flat, broad forehead, depressed in between the eyes. The animals are maintained in a semi-intensive management system, and rarely fed any concentrate.
- Lactating cows and bullocks are fed with straw, rice bran and kitchen waste. Cows are **milked only once in morning hours**. The milk yield per lactation ranges from 450 to 650 kg, with milk fat ranging from 4.8 to 4.9%.

27. Kenkatha

Fig. 39: Kenkatha breed: bull and cow

- Also known as "Kenwaria", the breeding tracts comprise of **Tikamgarh district of Madhya Pradesh and Lalitpur, Hamirpur and Banda districts of Uttar Pradesh.**
- Kenkatha breed is maintained mainly for draught purpose and are very popular for light draught on road and for cultivation.
- The animals are small, sturdy and powerful, varying in colour from grey on the barrel to dark grey on the rest of the body.
- The horns emerge from the outer angles of the poll in forward direction and end in sharp points.
- The animals are maintained in an extensive management system.
- Cows and young stock are maintained only on grazing while bullocks are usually fed good quality straws. They are known for their peculiar characteristic to thrive on poor feed resources.
- Kenkatha cows exhibit a daily milk yield ranging from 1.0 to 3.0 kg.
- During a lactation period of 6-8 months, the average production of Kenkatha cows ranged from 500-600 kg with milk fat percentage is 4.9.

28. Khariar

Fig. 40: Khariar breed: bull and cow

- **Khariar is named after its native tract "Khariar" in Nuapada district of Odisha**. Breeding tract comprised of **Nuapada, Kalahandi and Balangir districts of Odisha.**
- Heavy concentration of animals of this breed is found in Khariar, Komna, Sinapali and Boden blocks of Nuapada district.
- **Coat colours are mainly brown and sometimes grey.** Horns are straight and often emerge upward and inward. Animals of this breed are small sized and of strong build.
- Hump, neck, and some regions of the face and back are dark in colour.
- **The breed is used for draught purposes in its native tract, which is hilly and undulated. The lactation milk yield ranged from 300 to 450 kg with 4 to 5% fat.**

29. Kherigarh

Fig. 41: Kherigarh breed: bull and cow

- Kherigarh is a draught purpose breed and the breeding tract includes **"Kheri" district of Uttar Pradesh.** The breed has been named after this area.
- The breed is also known by various names like **"Kheri", "Kharigarh" and "Khari".**
- Bullocks are very good for draught purposes and can run very fast. The animals of this breed are small but active.
- Kherigarh has a white coat colour. **Some animals have grey colour distributed all over the body, especially on the face.**
- Horns are upstanding, curving outward and upward and thick at the base.
- The standard lactation milk yield ranges from 300 to 500 kg.

30. Konkan Kapila

Fig. 42: Konkan Kapila breed: bull and cow

- Konkan Kapila is a draught breed of Maharashtra, which is also known as "**Konkan gidda**" and "**Konkan**". The animals are useful to the farmers in all kinds of agricultural operations.
- The breeding tracts include **Thane, Raigadh, Ratnagiri, Sindhudurg and Palghar districts of Maharashtra.** In some parts of the **Konkan area in Maharashtra**, Konkan Kapila cows are the only source of milk due to non-availability of dairy milk because of lack of transport facilities.
- Animals have a small to medium sized and compact body, horizontal ears and straight forehead. Eyelids, muzzle, hoof and tail switch are generally black.

- Konkan Kapila cows are low milk producers (around 2.25 kg in day) and possess good draft ability suited to hilly terrain and hot and humid climate of its native tract and are significantly contributing to the livelihood of small and marginal farmers.
- Average milk yield of the animals is 450 kg with an average milk fat of 4.5 % (ranges from 3.5 to 5.7%).

31. Kosali

Fig. 43: Kosali breed: bull and cow

- The breed is predominantly seen in the **plains of Chhattisgarh** and the breeding tracts include **Raipur, Durg, Bilaspur and Janjgir districts**. The ancient name of this region was '**Kowshal**'.
- Farmers particularly of Yadava/Rawuth community are keeping this breed of cattle, generation after generation. **These are mainly reared for manure, draught and milk.**
- Most predominant colour is light red, followed by whitish grey. Few animals having black coat colour or red with white patches are also seen.
- Horns are stumpy, emerging straight, then going outward, upward, and inward from the polls. Muzzle, eyelids, tail switch and hooves are black. Head is broad, flat and straight.
- Hump is small to medium in size. **Udder small and bowl shaped. Bullocks of this breed are very efficient for ploughing in paddy fields and other operations in the paddy fields.**
- The average milk yield per lactation is 210 kg with an average milk fat percent of 3.5. The lactation yield ranged from 200 to 250 kg with average fat percent from 3 to 4.5.

32. Ladakhi

Fig. 44: Ladakhi breed: bull and cow

- Ladakhi is a draught breed of Jammu & Kashmir which originated from Ladakh region of Jammu and Kashmir. Leh and Kargil districts of Jammu and Kashmir are the main breeding tracts of this breed.
- Milk of this cattle serves as an important protein source for local people, particularly during lean winter period. Milk has a high fat percentage of around 5% and is used mainly for producing butter and churpi, an important part of the diet of local people.
- These are compact, small-sized and short-statured cattle with short limbs.
- They are mostly black-coloured cattle, but brown coloured ones are also seen.
- The horns are curved slightly, upward and forward, ending with pointed tips.
- The forehead is straight, small and hairy with a slightly long face.
- The udder is small in size and bowl-shaped.
- Body length of both male and female averages at 88 to 89 cm.
- Height of both males and females averages at 93 cm.
- Chest girth of both males and females averages between 117 to 119 cm.
- Parturition period averages to about 350 days.
- Age at first calving averages at 48 months.
- The milk yield is approximately 2 to 5 kg per day with milk fat average as 4.6%

- Milk is used mainly for producing butter and churpi, an important part of the diet of local people.
- This breed is well adapted to high altitude, extreme cold climate and hypoxic conditions. Body is compact with short legs that make it more adapted to mountainous terrains.
- Despite of extreme climatic conditions, subsistence on poor quality feed and low availability of water, it produces around 2 to 5 kg of milk per day. Animals are mainly used for milk, draught and manure.

33. Lakhimi

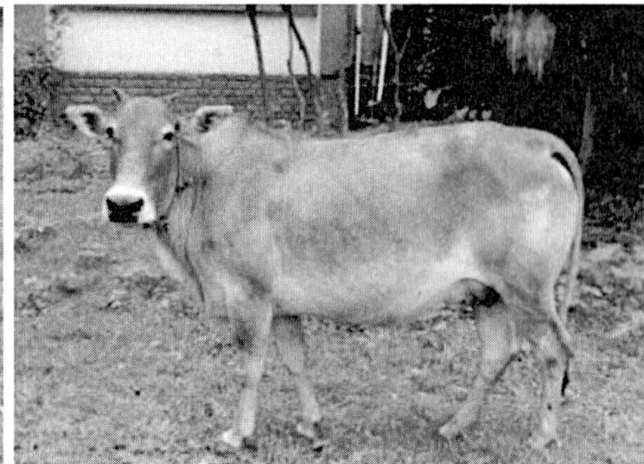

Fig. 45: Lakhimi breed: bull and cow

- Lakhimi is a dual-purpose breed of cattle found in the entire **state of Assam**. Animals are adapted to subtropical climatic conditions.
- **Bullocks are excellent draft animals for carting and ploughing especially in the muddy fields for paddy cultivation.**
- The **milk produced by Lakhimi cattle is rich in fat for which it fetches a high price**.
- Animals are found in brown and grey colour and are small sized, horned and have relatively short legs.
- Coat colour is variable mainly between brown and grey. Hump is medium in size and the backline is slightly curved.
- Udder is small and bowel shaped. Horns are straight. These cattle are maintained traditionally under a low/zero input system.
- Milking is done only once in the morning. The milk yield is low with an average of 359 kg per lactation (ranged between 325 to 375 kg per lactation) with an average milk fat 5.3% (ranging from 4.3 to 6.3 %).

34. Malnad Gidda

Fig. 46: Malnad Gidda breed: bull and cow

- Malnad Gidda is also known as "Gidda", "Uradana" and "Varshagandhi". "Malnad" means a hilly region and "Gidda" means small or dwarf. The breeding tracts of this breed include **Chikmagalur, Dakshin Kannada, Hassan, Kodagu, Shimoga, Uttar Kannada and Udupi districts of Karnataka.**
- This breed play a major role in the rural economy of this region by providing milk, manure and draft power with negligible inputs.
- The animals of this breed are well adapted to the local agro-ecological systems of Western Ghats. The predominant **coat colour is black with light shades of fawn on the thigh and shoulder region.**
- Horns are generally small, straight, outward, upward and inward. They are small with compact body frames and adult animals are around 90 cm tall.
- Tail switch is black in colour, the hump is small, the udder is also small and bowl shaped.
- Malnad Gidda cattle are reared under a low input low output system. In some areas, it is a zero input system where animals sustain solely on grazing.
- **Elite cows produce 3-5 kg of milk per day with a fat content of 5.5 to 8%** and the average lactation yield is around 220 kg. The animals of this breed are highly adapted to harsh climates, including heavy rainfall.

35. Malvi

Fig. 47: Malvi breed: bull and cow

- Malvi is named after its place of origin viz. "**Malwa**" region. It is also known by synonyms as "**Mahadeo puri**" and "**Manthani**". The breeding tracts include **Rajgarh, Shajapur, Ratlam and Ujjain districts of Madhya Pradesh**.
- Malvi cattle are white or white greyish - darker in males, with neck, shoulders, hump and quarters almost black. Cows and bullocks become nearly pure white with age.
- The horns are curved and emerge from the outer angle of the poll in an outward and upward direction and are about 20 - 25 cm in length.
- The animals are well known for quick transportation, endurance and ability to carry heavy loads on rough roads.
- The animals of this breed are strong and well–built. **The average milk yield of the cows of this breed is around 916 kg per lactation with 4.3% fat. The lactation milk yield ranged from 627 kg to 1227 kg.**

36. Mewati

Fig. 48: Mewati breed: bull and cow

- **Mewati, also known as "Kosi" or "Mehwati", is a draught purpose breed. The breeding tracts include Gurgaon and Faridabad districts of Haryana: Alwar and Bharatpur Districts of Rajasthan; and Mathura District of Uttar Pradesh.**
- Mewati cattle are powerful and docile, and are useful for heavy ploughing, carting and drawing water from deep wells.
- **The animal of this breed phenotypically similar to Hariana but there are also traces of influence of Gir, Kankrej and Malvi breeds.**
- Mewati cattle are usually white with a neck, shoulders and quarters of a darker shade. Horns are small to medium in size and emerge outwards, upwards and then inwards in the majority of animals.
- However, in some animals the horns emerge outwards and upward. Tips of the horns are pointed.
- Face is long and narrow with a straight, sometimes slightly bulging forehead.
- The average lactation yield reported is around 958 kg. The fat content in their milk varying from 4 to 4.5%.

37. Motu

Fig. 49: Motu breed: bull and cow

- This ia a dwarf breed of cattle which are used for draught purposes in hilly and undulated terrain. The breed is named after local area "Motu" of **Malkangiri district in Orissa.**
- Their breeding tract comprises southern part of Malkangiri district and adjoining areas of **Chhattisgarh and Andhra Pradesh.**
- Heavy concentration of animals is found in Motu, Kalimela, Podia and Malkangiri areas of Malkangiri district in Orissa.
- Sandy and clay type soil is predominant in this area. Coat colour is mainly brown (reddish) and sometimes grey. Few animals are white in colour.
- Animals of this breed are mostly polled and horns whenever present emerge straight and upward and end with rounded tip.
- Though the breed is small sized they are strong in built. The milk yield is meagre and ranged from 100 to 140 Kg in a lactation with 4.8 to 5.3% fat.

38. Nagori

- The breeding tracts of Nagori breed are **Bikaner, Jodhpur and Nagaur district of Rajasthan.** The breed takes its name from the name of the home tract i.e., Nagaur district but animals are also found in adjoining Jodhpur district and Nokha tehsil of Bikaner district.
- They are white in colour and are upstanding, very alert and agile animals with long and narrow face like that of a horse.
- Considering the proximity of the native tracts of the **Hariana breed** of cattle in the north and northeast and **Kankrej breed** of cattle in the south

and southwest, it seems reasonable to suppose that **Nagori cattle may have evolved from these breeds.**

- Nagori cattle are fine, large, upstanding, active and docile, with white and grey colour. They have long, deep and powerful frames, with straight backs and well-developed quarters.

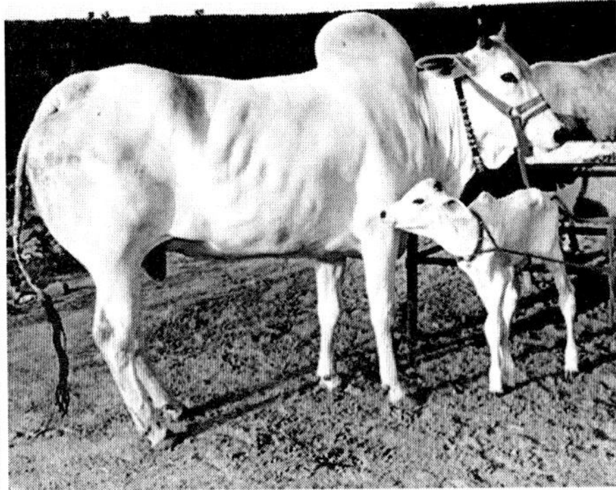

Fig. 50: Nagori breed: bull and cow

- The face is long and narrow but the forehead is flat and not so prominent. The eyelids are rather heavy and overhanging and the eyes are small, clear and bright.
- The ears are large and pendulous. The horns are moderate in size, emerge from the outer angles of the poll in an outward direction, and are carried upwards with a gentle curve to turn in at the points.
- **The Nagori breed is one of the most famous trotting draughts breeds of India and are generally appreciated for fast draught activity.** They are famous as trotters and are used all over Rajputana of Rajasthan in light iron-wheeled carts for quick transportation.
- **Average milk yield per lactation of Nagori cattle is 603 kg with an average milk fat of 5.8%. The lactation yield ranged from 479 to 905 kg.** Present population size of Nagori cattle is 3,47,081 as per 20th Livestock Census, 2019.

39. Nimari

- Nimari, also known as "Khargaon", "Khargoni" and "Khurgoni", is a prominent draught breed of central India named after its place of origin (**Nimar region of Madhya Pradesh**).
- Its breeding tracts comprised of Khargaon (West Nimar) and Badwani districts of Madhya Pradesh.

Fig. 51: Nimari breed: bull and cow

- Few purebred animals are also found in adjoining districts of Maharashtra. Believed to have **originated from crossing of Gir and Khillari breeds**, having inherit characteristics of both these breeds.
- It is red in colour with large white splashes and possesses a massive body, convex forehead, and has horns that resembles Gir breed.
- Horns usually emerge in a backward direction from the outer angles of the poll, somewhat in the same manner as in Gir cattle, turning upward, outward and finally inward. Hardiness, agility and temper have been inherited from Khillari.
- Prized primarily for draught work, the animals are maintained chiefly for agricultural operations with occasional use in transportation.
- Average milk yield of this breed is 767 kg per lactation with a range of 600-954 kg per lactation. The average milk fat is 4.9%. Present population size of Nimari cattle is 4,79,061 as per 20th Livestock Census, 2019.

40. Ponwar

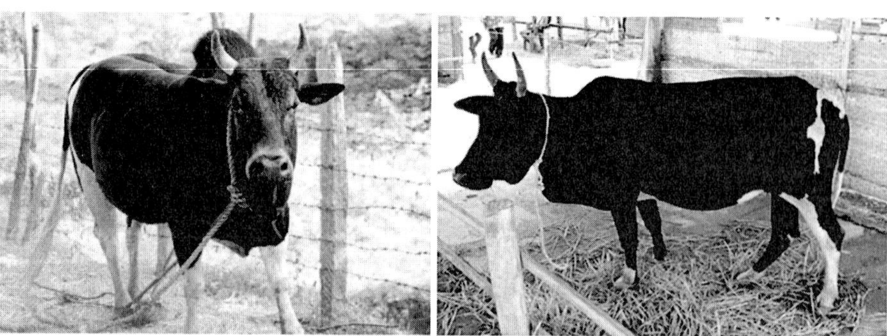

Fig. 52: Ponwar breed: bull and cow

- Ponwar is an indigenous cattle breed of India. It is known to have originated at Ponwar in Puranpur Taluk in Pilibhit district of Uttar Pradesh state and the breed is named after the same place. The cattle is also restricted to the small geographic area adjoining Ponwar.
- Ponwar, a draught breed of cattle from North India, is believed to have evolved from a **mix of hilly cattle (Morang - Nepalese hill cattle) and plain land cattle. It is also known as "Purnea"**. Its breeding tract is the **Pilibhit district of Uttar Pradesh.**
- There is no particular pattern but black and white patches are intermixed.
- They possess medium sized horns, which generally emerge outward, upward and then curve inward with pointed tips.
- The animals behave in a semi-wild manner and are tough to control. They move in groups putting their heads down in between each other which is a peculiar behaviour that has emerged due to fear from predators.
- The animals are maintained in an extensive management system, without provision of sheds and are raised entirely on grazing in forestland without feed supplementation.
- Females are generally not milked and calves are allowed to suckle the milk. Bullocks are fast movers and therefore used both for agricultural operations and transportation.
- **Cows produce a fair amount of milk with an average yield of 459 kg per lactation.** The milk produced is rich with **about 4.3% fat content.** Present population size of Ponwar cattle 28,656 is as per 20th Livestock Census, 2019.

41. Red Kandhari

Fig. 53: Red Kandhari breed: bull and cow

- Originated from **Kandhar tehsil in Nanded district of Maharashtra, Red Kandhari is also known as "Lakhalbunda". It is a draught breed of cattle.**
- Its breeding tracts comprised of Ahmadnagar, Parbhani, Beed, Nanded and Latur districts of Maharashtra. The colour is uniform deep dark red, but variations from a dull red to almost brown are also found. Bulls are a shade darker than cows.
- Horns are evenly curved and medium sized. The animals are maintained under an extensive management system on grazing only in small herds.
- Small amount of concentrate is offered to bullocks, male calves and milking animals. The bullocks are used for heavy agricultural work like ploughing and carting as well as for transportation.
- **The cows produce a fair quantity of milk with an average of 598 kg per lactation with average fat percentage of 4.57.** Present population size of Red Kandhari cattle is 1,49,221 as per 20th Livestock Census, 2019.

42. Siri

Fig. 54: Siri breed: bull and cow

- Siri is a small sized draught purpose breed of hilly region of **West Bengal and Sikkim**. The breed is also known as "Trahbum". The breeding tracts of the breed include **Darjeeling district of West Bengal and Sikkim north, Gyalshing, Namchi, Gangtok, Sikkim east, Sikkim south and Sikkim west.**
- The breed is either black or brown with white patches, though totally black or brown animals are also available. The hump is cervico-thoracic with tufts of hair on it.

- Horns are medium sized and curved outward, forward and slightly upward with sharp and pointed tips.
- The animals can graze in the steep slope of a hilly forest. Animals are housed on slopes of hills in open houses.
- The males are mainly reared for draught purposes in the hilly area and sometimes they are the only source of draught power.
- The breed produces very less amount of milk averaging around 425.8 kg per lactation with fat percentage varying from 2.8 to 5.5.

43. Vechur

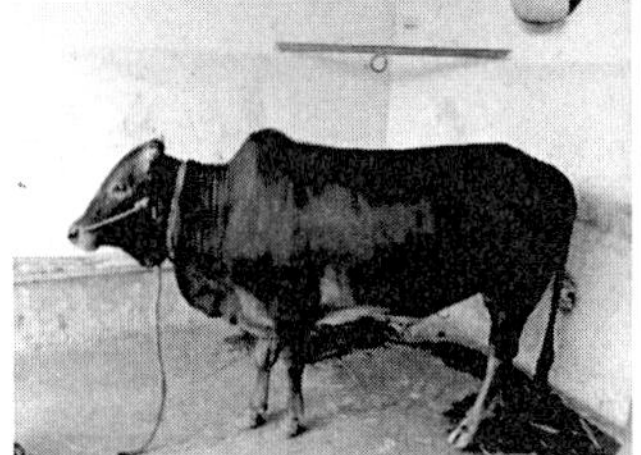
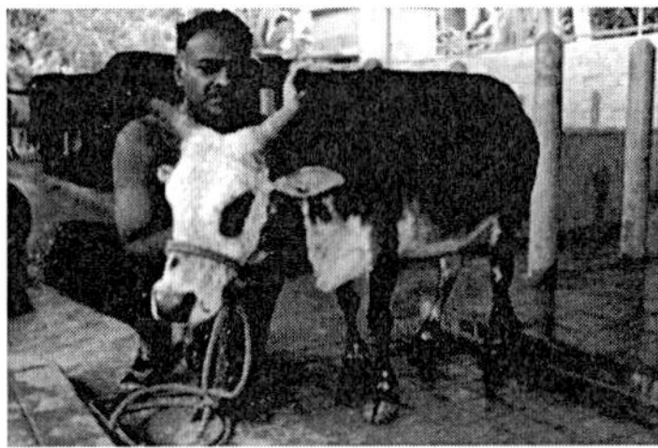

Fig. 55. Vechur breed: bull and cow

- Vechur is one of the **dwarf cattle breeds** of India, with an average length of 124 cm and height of 87 cm, it is considered to be the **smallest cattle breed in the World**.
- According to the **Guinness book of records**, this is the smallest cattle breed in the World.
- **It is known by the name of a place Vechur, a small place by the side of Vembanad lake near Vaikam in Kottayam district of South Kerala.**
- **The breeding tracts include Alapuzzha/ Alleppey, Kottayam, Pathanamthitta and Kasargode districts of Kerala.**
- The animals are light red, black or fawn and white in colour. In bulls, colour in between fore and hindquarters is relatively dark or dark grey.
- Horns are small, thin curving forward and downward. In some cases, they are extremely small and are hardly visible.
- The animals are well adapted to the hot and humid climate of the area.

- The animals are maintained for manure and milk. Milk production is relatively higher than any other dwarf cattle.
- **Average milk yield is 561 kg per lactation and the milk fat percent ranged from 4.7 to 5.8.**
- The population size of this breed is 15,181 as per 20th Livestock census, 2019.

44. Punganur

- Punganur is a **miniature dual purpose cattle breed** and mainly confined to **Chittoor district of Andhra Pradesh.** This breed was developed **by Rulers of Punganur area** and hence named after the area.
- The breeding tracts are confined to the taluks of Punganur and adjacent taluks of **Vayalpad, Madnapalli and Palamaner in Chittoor district of Andhra Pradesh.**
- It is mainly maintained at **Livestock Research Station, Palamaner, Chittoor district, Andhra Pradesh, attached to Shri Venkateswara Veterinary University, Tirupati.**
- Some animals are also maintained by private breeders. It is white, grey or light brown to dark brown or red in colour. Sometimes, animals with white colour mixed with red, brown or black coloured patches are also seen.
- Well-noticed by its **short stature, the breed has a broad forehead and short horns**. The horns are crescent shaped and often loose curving backward and forward in males and lateral and forward in females.
- Punganur cows, known for more than just their petite size, have become legendary within Andhra Pradesh, India. Notably regarded for holding the Guinness World Record of being the smallest cow breed (at 97-107 cm tall and 115-200 kg, respectively).
- Known for draught resistance, it can thrive well on dry fodder feeding. The bullocks are used for agricultural operation in light soil as well as for driving carts for transportation and special races.
- **Average milk yield of the breed is 546 kg per lactation (ranging from 194 to 1100 kg) with 5% average milk fat.**
- The population size of this breed is 13,275 as per 20th Livestock census, 2019.

Fig. 56: Punganur breed: bull and cow

45. Poda Thurpu

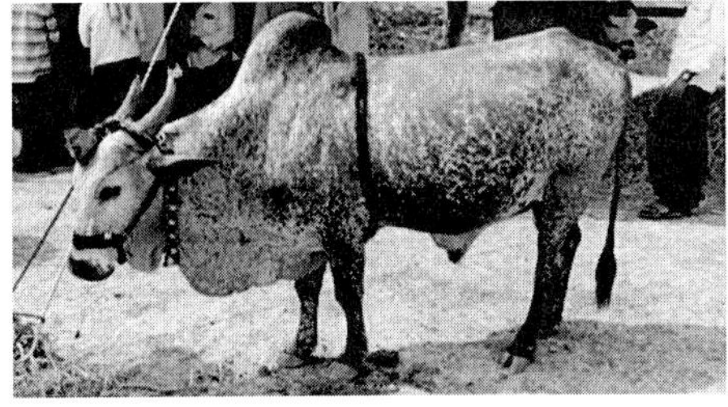

Fig. 57: Poda Thurpu breed: bull and cow

- These are medium-sized cattle with compact bodies distributed in **Nagarkurnool district of Telangana.**
- Animals have **white coats with brown patches or red/brown coats with white patches.**
- Horns are straight and upward or forward pointing in orientation, and broad at the base. Forehead is convex with a deep groove at the centre in the majority of cases.
- These animals are usually maintained under an open grazing system and have excellent long migration capacity over undulated forest topography.
- Poda Thurpu cattle are reared mainly for draught purposes. Bullocks have excellent draught capacity in terms of endurance, speed & stamina and are preferred in both dryland and wetland agriculture.
- **Daily milk yield ranged from 2 to 3 kg and lactation milk yield varied from 494 to 646 kg.**

- **Herd size ranged from 23 to 75. Population size is approximately 15,000.**

46. Nari

- **Nari cattle are native to Sirohi and Pali districts of Rajasthan, Sabarkantha and Banaskantha districts of Gujarat.**
- These are medium in size with excellent migratory capacity and can survive well on grazing, and in the open housing system during all kinds of weather.
- Coat colours are white or greyish white in the majority of cows and white, greyish white and sometimes black in bulls.
- **Horns are spirally curved and outward /forward pointing in orientation.**
- Forehead is broad and slightly concave in the majority of cases. These are dual purpose cattle used for both milk and draught purposes.
- **Daily milk yield ranged from 5 to 9 kg and lactation milk yield ranged from 1119 to 2223 kg.**
- **Excellent in draught power in both plains and hilly forest areas. Population size is approximately 55,000.**

Nari bull

Nari cow

Fig. 58: Nari breed: bull and cow

47. Dagri

- Dagri cattle are distributed in **Dahod, Chhotaudepur and parts of Mahisagar, Panchmahals and Narmada district of Gujarat.**
- Coat colours are predominantly white, sometimes with grey shade. Small sized animal with a compact body and straight forehead.

- Horns are curved upward in a lyre shape or straight with pointed tips.
- Extensively used as draught purpose animal for agricultural operations in hilly areas.
- **The cows of this breed produce less milk i.e., 1.5-3.0 kg/day and 75-650 kg/lactation, which is mainly used for household consumption.**
- **The animals of this breed have lesser feed requirement, survives mainly on grazing, hardly stall fed.**
- **The population size is approximately 2,80,000 heads.**

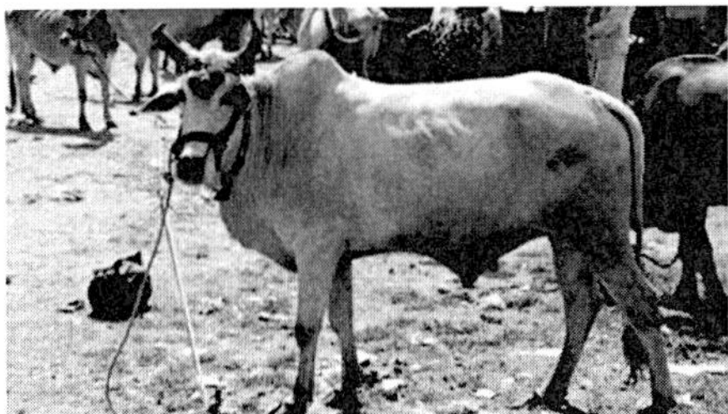

Fig. 59: Dagri breed: bull and cow

48. Thutho

Fig. 60: Thutho breed: bull and cow

- Thutho breed of cattle is also known as "Ameshi", "Sheapi", "Chokru" and "Tseso". Thutho cattle are available in all districts of Nagaland.
- This breed is mainly used for meat, draught and manure.
- Animals are well adapted to hilly regions and able to graze on hill slopes even during rainfall. The colour of the animals are black or brown, sometimes white patches on the face and body.

- **Animals are medium in size, hardy, well-built and docile. Fore-head is small and straight. Backline is uneven, slopes behind the small hump, rises to peak between hip bones and then drops sharply to the tail head. Horns are curved outward and upward. Short and stumpy.**
- The animals are maintained in an extensive management system and generally maintained on grazing only. Cows are generally not milked.

49. Shweta Kapila

- Complete white coloured cattle found in **North Goa and South Goa districts of Goa State**.
- White colour extends from muzzle to tail switch including eyelashes and muzzle (whitish brown).
- Short to medium statured animal with straight face, straight and small horns directed upward and outward, and small to medium hump.
- Udder is bowl shaped and small to medium in size with cylindrical teats having rounded tips. Height ranges from 97 to 137 cm.
- **Daily milk yield ranged from 1.8 to 3.4 kg with an average of 2.8 kg and lactation milk yield varied from 250 to 650 kg with an average milk fat of 5.21% (ranged from 4.5 to 6.4%).**
- **Population size is approximately 22,000 heads.**

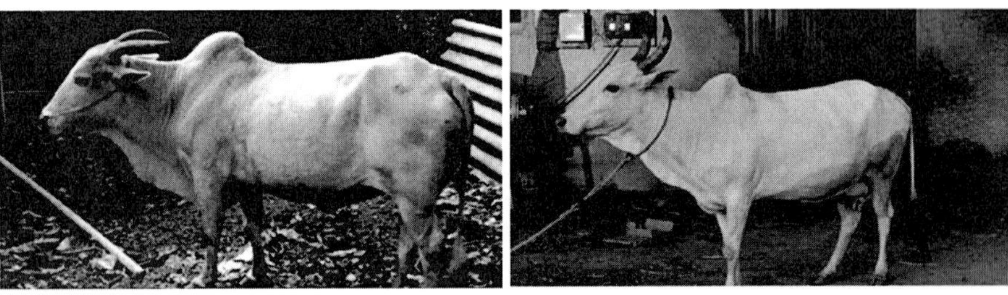

Fig. 61: Shweta Kapila breed: bull and cow

50. Himachali Pahari

- Primary cattle population distributed in **mid-hill to high hills region of Himachal Pradesh (Kullu, Chamba, Mandi, Kangra, Sirmaur, Shimla, Kinnaur, Lahul Spiti districts) and adapted to mountain topography,** extremely cold climate, and fodder scarcity.

- **Small to medium sized animal, primarily black and blackish brown in colour** with compact cylindrical body, short legs, medium hump, horizontally placed ears and comparatively long tail.
- Horns are medium in size and are mainly curved in lateral and upward direction.
- Well suited to mountainous livestock production systems as a source of **milk, draught and manure.**
- Adult weight varies from 200 to 280 kg in males and 140 to 230 kg in females.
- Height varies from 90 to 120 cm.
- **Daily milk yield ranges from 1 to 3 kg and lactation milk yield varied from 350 to 650 kg with an average milk fat of 4.68% (ranged from 4.06 to 5.83%).**
- Population size is approximately 7,59,000 heads.

Fig. 62: Himachali Pahari breed: bull and cow

51. Purnea

- Purnea cattle are distributed in **Araria, Purnea and Katihar districts; and the Adjoining areas of Kishanganj, Supaul and Madhepura districts of Bihar.**
- These are small sized animals with three different coat colours viz. grey, red and black.
- These are reared for milk, manure and to some extent for draught power. These cattle have medium hump, small to medium dewlap, small naval flap and small to medium sized udder.

- Head is medium in size. Most of the Purnea cattle have flat poll while some have moderately prominent poll. Horns are small in size, straight and mostly carried upwards sometimes laterally.
- **Daily milk yield ranged from 1 to 5 kg and lactation milk yield from 360 to 785 kg with an average milk fat of 4.22 % (ranged from 4 to 4.5%).**
- **Population size is approximately 2,19,000 heads.**

Fig. 63: Purnea breed: bull and cow

52. Kathani

- It is a dual purpose, medium sized cattle with a compact body, horizontal ears, and straight forehead.
- It is distributed in **Chandrapur, Gadchiroli and Gondia districts of Western Maharashtra (Vidarbha region).**
- Kathani cattle are predominantly **white, reddish, and blackish in colour**.
- Horns are straight and curved, poll non-prominent and dewlap small to medium in size.
- Kathani cattle are well adapted in a low input production system and possess good draft ability suited to marshy land for paddy cultivation in stagnated water and rains without having any hoof trouble.
- Average daily milk yield, lactation milk yield and lactation length are 0.55±0.01 kg, 193.07±5.28 kg and 237.76±1.82 days, respectively.

Fig. 64: Kathani breed: bull and cow

53. Sanchori

- It is medium sized cattle with the majority of animals are predominantly white in colour with large dewlap.
- It is distributed in **Sanchore, Raniwara, Bhinmal, Bagoda and Chitalwana Blocks of Jalore district of Rajasthan.**
- Horns are curved and outward upward and inward pointing in orientation. Face is moderate in length and the forehead is fairly broader and slightly concave.
- Sanchori cattle are good milk producers. Average daily milk yield, lactation milk yield and lactation length are 9.08±0.16 kg, 2769.40±48.80 kg and 9.88±0.14 months, respectively.
- The Sanchori cattle exhibit a fat percentage ranging from 2.9 to 5.1%, with an average of 4.04%.
- The population size of Sanchori cattle is relatively small, numbering in the few thousands.

Fig. 65: Sanchori breed: bull and cow

54. Masilum

- Masilum cattle are **small in size, well built, sturdy, and well adapted** to the hill ecosystem of Meghalaya.
- These indigenous cattle are available in Hills of Meghalaya and reared by the **Khasi and Janitia community**.
- The Khasi language has words '**Masi' and "Lum" that means cattle and hills, so it is called "Masilum**".
- The predominant body colour varied from black, brown, and mixture of brown, grey and black.
- Dewlap and hump are medium in cow while well-developed dewlap and hump with tuft of hair over the hump has been observed in bulls.
- Horns are short and black in colour.
- Average daily milk yield, lactation milk yield and lactation length are 2.72±0.45 kg, 456.42±10.53 kg and 168.56±9.28 days, respectively.

Fig. 66: Masilum breed: bull and cow

Crossbred cattle in India

The crossbreeding of Indian breeds with exotic was started in India through All India Co-ordinated Research Project (AICRP) on cattle in the year 1968 to enhance the milk production of Indian cattle breeds.

The AICRP on cattle involved the crossing of indigenous cattle breeds (Hariana, Ongole, Gir, Tharparkar, Sahiwal, Red Sindhi and Local) with superior exotic breeds (HF, Jersey, Brown Swiss). The objective of AICRP on cattle was to know the optimum combination and level of exotic inheritance based on assessing the production and reproduction performance of the cross breeds in different agro-climatic regions. A control frozen semen bank was also established at Hesarghatta (Karnataka) for collection, processing, storage and distribution of frozen semen on the basis of the results obtained it was found that crossbreds with ¾ exotic inheritance did not do better than half breds due to disease problem, availability of required input and management problems etc. The better performance had been of those having 50 to 62.50% level of exotic inheritance than all other crossbreds.

According to the 20th National Livestock Census (2019), the crossbred female population is 50.42 million and there is a huge gap in the number of crossbred bulls available and actually required to breed this vast female cattle population in the country. The Frieswal project is the only project in the country where more than 250 bulls progeny tested were produced. The quality semen produced from bulls borne out by mating of proven bulls and elite cows can be utilized in the field conditions to improve the production potential of the crossbred population. Also, the proven bull semen is also made available to the stakeholders for artificial insemination of the elite animals in the progressive farmers herd. Further per year more than 100 male calves are produced by elite mating, under the project. These male calves can be extensively used for semen collection and genetic improvement programmes of the country.

1. Taylor breed: European missionaries started crossbreeding of cattle in India in 1875 when the **Taylor breed** of cattle was developed around Patna, Bihar, by crossing of **Shorthorn bulls with native cows**. This breed was evolved from a crossbred population in and around Patna. It originated in crosses between local cattle and four Shorthorn (or perhaps Channel Island) bulls introduced in 1856 by the Commissioner William Taylor. No conscious selection was made among crossbreds but they were better milkers (6 to 8 kg milk / day) than the local cattle. These animals are black, grey or red in colour.

Fig. 67: Taylor crossbreed cow

2. Santa Gertrudis, a breed of beef cattle developed in the 20th century by the King Ranch in Texas. It originally resulted from crossing **Brahman bulls of about seven-eighths pure breeding and purebred Shorthorn cows.** Over a period of years beginning with first crosses in 1910, selective breeding was practiced in which preference was given to red colour without sacrificing type and conformation. Santa Gertrudis cattle are usually solid red in colour with occasional small white markings on the forehead or in the region of the flanks. They were developed from foundation stocks of about five-eighths Shorthorn and three-eighths Brahman ancestry.

Santa Gertrudis cattle are generally the heaviest of the beef breeds. They have great depth and length of body, with more loose skin about the neck, brisket, and navel than the breeds of strictly British origin. They proved to be highly adaptable to the semitropical U.S. Gulf Coast.

Fig. 68: Santa Gertrudis crossbreed bull

3. Jersind: The crossbreeding policy was modified in 1953 and the Jersey bulls were used on Red Sindhi cows and on cows with 7/8 or more Red Sindhi inheritance to evolve a crossbred, to be called '**Jersind**' having between 3/8 to 5/8 Jersey inheritances.

The half-bred Jersey-Red Sindhi cows calved earlier (29 months) and also produced more milk in the first 305 days of their first lactation (1968 kg) than the Red Sindhi cows. They also exceeded Red Sindhi cows in the total life production by 2.09 times. The average lactation yield of Jersind cows declined to 1437 kg in later generations. The breed has shown deterioration over the years mainly because of small numbers and is only confined to the farm.

Fig. 69: Jersind crossbreed cow

4. Karan Swiss: Another high milk producing strains, namely **Karan Swiss**, have been developed at ICAR-National Dairy Research Institute (ICAR-NDRI), Karnal (Haryana), India through extensive experiments of crossbreeding. The highest milk yield recorded in the case of Karan Fries was 41.8 kg and in Karan Swiss, it touched 30 kg.

The **Karan Swiss** is a dual-purpose breed with the oxen being well suited for work and the cows giving good quantities of milk. High producing females will produce **5000 to 6000 kg with a 4.78% butterfat** during a lactation. The breed is usually light grey to dark brown in colour.

Fig. 70: Karan Swiss crossbreed cow

5. The **Karan Fries** were developed in India at the National Dairy Research Institute at Karnal. The breed was developed using **Holstein (Friesian) and Tharparkar**. The percentage of Holstein in the breed ranges from 3/8 to 1/2 of the breeding.

This crossbreed cattle was established in the 1980s at the ICAR-NDRI and Karan-Fries cattle are known for their high milk production. On average, they yield around 3700 kg of milk with a fat content of 3.8% to 4%.

Fig. 71: Karan Fries crossbreed cow

6. Sunandini is a composite breed of cattle developed in Kerala by crossing **nondescript cattle with Brown Swiss initially, later on Jersey and Holstein Friesian inheritance was added in the plan.** Jersey and HF bulls from many parts of the world are extensively utilized to generate this breed.

The breed characteristics fixed for the cows are 350-400 kg mature body weight, 28-32 months age at 1st calving, **2300-2700 kg 1st lactation milk yield, 3200 kg overall lactation yield and 4% milk fat.** The present breeding policy for Sunandini aimed at creating a new synthetic breed of a crossbred population with **exotic inheritance of around 50% from Jersey, Brown Swiss and Holstein.**

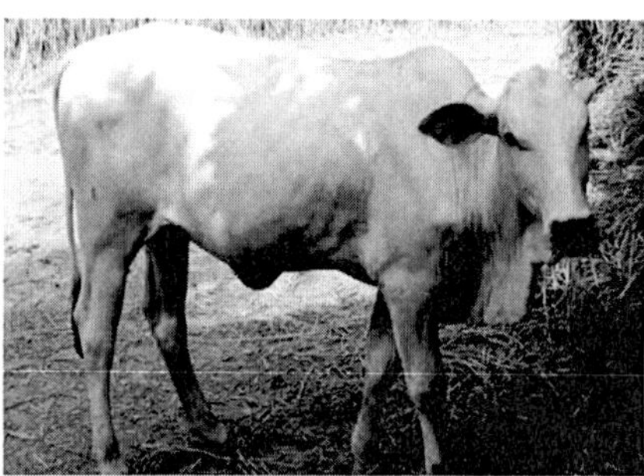

Fig. 72: Sunandini crossbreed cow

7. Brownsind: In 1955, since Brown Swiss was found to be most heat tolerant, it was used instead of Jersey for crossbreeding on pure Red Sindhi cows and on cows with 7/8 or more Red Sindhi inheritance to evolve a crossbred, to be called '**Brownsind**' having between 3/8 to 5/8 Brown Swiss inheritance.

8. Hardhenu

- This breed was developed at Cattle Breeding Farm (CBF), Lala Lajpat Rai University of Veterinary and Animal Sciences (LUVAS), Hisar over a period of 20 years from 1997 to 2016.
- Hardhenu is a remarkable crossbreed, resulting from a fusion of North American Holstein Friesian, Hariana, and Sahiwal breeds. Its genetic makeup comprises **62.5% exotic (from the Holstein Friesian)** and **37.5% indigenous** (from the **Hariana and Sahiwal**) components. Let's delve into some fascinating aspects of this crossbreed:
- Milk Production: Hardhenu is a boon for the dairy industry. These cows yield approximately 55 litres of milk per day, making them highly productive.
- The economic traits studied were first lactation milk yield (FLY; kg), first lactation length (FLL; days), first lactation peak yield (FPY; kg), first dry period (FDP; days), milk yield per day of calving interval (MCI; kg), milk yield per day of age at second calving (MSC; kg) and milk yield per day of lactation length (MLL; kg). The least square means for FLY, FLL, FPY, FDP, MCI, MSC and MLL were 2262.98±57.52 kg, 307.27±5.93 days, 10.30±0.21 kg, 110.97±6.47 days, 5.51±0.13 kg, 1.37±0.03 kg and 7.42±0.14 kg, respectively.
- Genetic and Phenotypic Correlations: Studies have revealed strong correlations among various production performance traits in Hardhenu crossbred cattle. For instance, factors like first-service period (FSP), first-calving interval (FCI), and calving-to-first-service interval (CFSI) are interconnected and can be improved through selective breeding.
- Improvement Over Time: Remarkably, all the production performance traits studied have shown consistent improvement over periods. This positive trend can be attributed to better selection, improved management practices, and nutrition at the farm.
- In summary, Hardhenu stands as a testament to the synergy between diverse breeds, resulting in a productive and adaptable crossbreed that contributes significantly to the dairy sector.

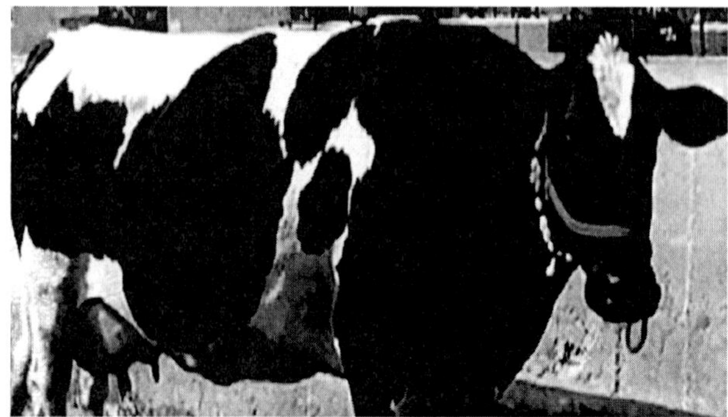

Fig. 73: Hardhenu crossbreed cow

9. Phule Triveni

- The 'Phule Triveni' is a new synthetic crossbred evolved at Mahatma Phule Krishi Vidyapeeth (MPKV), Rahuri, Maharashtra. The breed has been developed through crossing of Holstein Friesian, Jersey and Gir.
- The cow produces on an average 3000 to 3500 litre of milk per lactation with average fat content of 4.0 %.
- Developed by crossing of **Holstein Friesian (50%), Jersey (25%) and Gir (25%)** breeds.

Fig. 74: Phule Triveni crossbreed cow

10. The NBAGR also registered Frieswal cattle as the first synthetic breed in the country (December 2023). The **Frieswal cattle** is developed utilizing the crossbreds of (Holstein Friesian×Sahiwal) in 37 Military Farms to cover all the agro-climatic zones of the country. A Frieswal crossbred has 3/8 of

Sahiwal and 5/8 of Holstein-Friesian inheritance. This new breed has 62.5 percent exotic and 37.5 percent indigenous blood and is expected to yield approximately 4000 kg of milk, with 4% butterfat, in a lactation period of 300 days under good managemental conditions.

Fig. 75: Frieswal crossbreed bull and cow

Frieswal cattle is synthetic dairy cattle with Sahiwal (37.5) and Holstein Friesian (62.5) inheritance, developed by ICAR-Central Institute for Research on Cattle, Meerut (Uttar Pradesh), India. It has production potential of about 7000 kg of milk yield in a standard lactation with peak yield of about 41 kg. This breed is acclimatized to all agro-climatic regions of the country.

2

Buffalo Breeds of India

The buffalo belonged to the Bovidae family, *Bovinae* subfamily and the scientific name is ***Bubalus bubalis***. This is descended from the ***Bubalus arnee* or wild Indian buffalo**, and is widely dispersed throughout Southern Asia. The world population of buffaloes is approximately 204 million head: more than 98% are in Asia; 0.8%are in Africa, particularly in Egypt; 0.9% are in South America; and 0.2% are in Europe (FAO, 2024).

The countries with the largest numbers of dairy buffaloes are India, Pakistan, China, Nepal and Egypt. In Pakistan and Nepal there are more dairy buffaloes than dairy cows. Water buffaloes are the main source of milk in South Asia. India is the largest water buffalo milk producer compared to Pakistan, where buffaloes produce more milk than cattle.

Scientific classification

Genus: *Bubalus*

Species: *bubalis*

Wild water buffalo: *Bubalus arnee*

Domestic water buffalo: *Bubalus bubalis*

Lowland anoa: *Bubalus depressicornis*

Mountain anoa: *Bubalus quarlesi*

Tamaraw: *Bubalus mindorensis*

Cebu tamaraw: *Bubalus cebuensis*

The domestic water buffalo Bubalus bubalis, belonged to the family *Bovidae*, sub-family *Bovinae*, genus *bubalus* and species *arnee* or wild Indian buffaloes. Buffaloes are classified into two distinct classes: swamp buffalo and river buffalo.

Water buffaloes are the principal source of milk in South Asia

The largest water buffalo milk producers are India and Pakistan, where buffaloes produce more milk than cattle.

More than 95 % of the world's buffalo population is found in Asia where they play a leading role in the rural livestock production system. Over the past decade buffalo farming has widely expanded in Mediterranean areas and in Latin American countries and several herds have also been introduced in Central and Northern Europe.

The water buffalo (*Bubalus bubalis*) is a large bovine species originating in Asia; not to be confused with the **wild African buffalo (*Syncerus caffer*)**. Whilst their distribution is still concentrated largely in Asian countries, but small populations are also existed in other parts of the world. The vast majority of buffaloes are raised in labour-intensive, small family farms, where they provide milk and serve important functions as draft animals for agricultural practices. Buffalo milk has high fat content besides the important nutritional value it has twice the calorific content of cow's milk, extremely rich in calcium and is a good source of minerals like magnesium, potassium and phosphorus.

Buffalo farming is increasing in Italy too, due to the growing market demand for buffalo milk that is utilized exclusively for the production of "mozzarella cheese".

Another economic benefit derived from buffalo milk production, the milk of buffalo is not restricted by the specific European Union (EU) directive called "milk quotas", introduced to stop the increase of cow milk production. In fact, this regulation induced some farmers, in areas where Friesian cattle are traditionally reared, to consider the option of breeding milking buffaloes for the production of "mozzarella cheese".

The domestic **water buffalo (*Bubalus bubalis*)** contributes a significant share of global milk production and is the major milk producing animal in several countries. Buffaloes are kept mostly by small-scale producers in developing countries, who raise one or two animals in mixed crop livestock systems. **Water buffaloes are classified into two subspecies: the river buffalo and the swamp buffalo. River buffaloes** constitute approximately 70 percent of the world water buffalo population. River buffalo milk accounts for a substantial share of total milk production in India and Pakistan and is also important in the Near East. **Swamp buffaloes** are smaller and have lower milk yields than river buffaloes. They are found mainly in Eastern Asia and are primarily raised for draught power (FAO).

River buffaloes usually produce between 1500 and 4500 liters of milk per lactation. They have a significantly longer productive life span than cattle, providing calves and milk until they are up to 20 years of age. The factors that constrain the commercial buffalo milk production include animals' late age at first calving, the seasonality of oestrus, long calving interval and dry period.

In recent decades, breeding programmes especially in Bulgaria, China, Egypt, India and Pakistan, have attempted to improve the milk yield of river buffalo. Well-known specialized dairy buffalo breeds include Murrah, Nili-Ravi, Kundi, Surti, Jaffarabadi, Bhadawari and Mehsana (FAO).

Breeds of Indian buffalo

- India has to total 20 registered buffalo breeds till now.
- buffalo population 109.85 million (2019)

India is producing 24%, which is the highest milk yield (221.1 million tons) in the World, where **buffalo share 49% total milk yield production (major contributor than cow).**

- India produces **43% of the World buffalo meat production**.
- **Karyotype of buffalo**
    - Buffalo breeds are classified as Riverine type (or) Water buffaloes and Swamp type. In India, Riverine buffaloes are found which have different chromosome numbers (50) than the swamp buffalo (48).
    - The river buffalo karyotype (2n = 50) has five biarmed pairs (submetacentric chromosomes). The remaining chromosomes are acrocentric, including both the X, being the largest acrocentric chromosome, and Y, the latter being one of the smallest acrocentric chromosomes.
    - Data indicates that the bubaline lineage had frequent events of centric fusions and Robertsonian translocations, resulting in the formation of biarmed chromosomes.
    - The second largest pair of the Tamaraw buffalo originated from the Robertsonian translocation of telocentric pairs 7 and 15 of the river buffalo.

Buffalo can be classified as under following categories in India

- **Murrah Group** - Murrah, Nili-Ravi and Kundi
- **Gujarat Breeds** - Surti, Mehsana and Jaffarabadi

- **U.P. breeds** - Bhadawari and Tarai
- **Central Indian varieties** - Nagpuri, Pandharpuri, Manda, Jerangi, Kalahandi and Sambalpur
- **South Indian breeds** - Toda and South Kanara

Table 1: Indian breeds of buffalo

Sl. No.	Breed	Breeding Tract	Main Uses
1	Murrah	Haryana and Delhi	Milk
2	Mehsana	Gujarat	Milk
3	Nili-Ravi	Punjab	Milk
4	Pandharpuri	Maharashtra	Milk
5	Banni/Kundi	Gujarat	Milk
6	Surti	Gujarat	Milk
7	Dharwadi	Karnataka	Milk
8	Bhadawari	Uttar Pradesh and Madhya Pradesh	Milk and draught
9	Jaffarabadi	Gujarat	Milk and draught
10	Marathwadi	Maharashtra	Milk and draught
11	Purnathadi	Maharashtra	Milk and draught
12	Nagpuri	Maharashtra	Milk and draught
13	Gojri	Himachal Pradesh and Punjab	Milk and draught
14	Luit (Swamp)	Assam	Milk and draught
15	Chilika	Odisha	Milk, draught and manure
16	Manda	Odisha	Milk, draught and manure
17	Kalahandi	Odisha	Milk, draught, manure and horns are used for making handicrafts and house hold items
18	Toda	Tamil Nadu	Milk, draught, rituals, Socio-cultural, Religious ceremonies
19	Chhattisgarhi	Chhattisgarh.	Milk, draught and meat
20	Bargur	Tamil Nadu	Milk, manure and meat

Jerangi, Tarai and Godavari buffalo are unlisted by NBAGR till date.

Description of buffalo breeds

1. Murrah

- It is the best milch breed among all Indian breeds of buffaloes.
- Most important breed of buffaloes whose home tract is Rohtak, Hisar and Sind of Haryana, Nabha and Patiala districts of Punjab and southern parts of Delhi state. They are also found in Western Uttar Pradesh.

- "Murrah" breed is also known as **"Delhi", "Kundi" and "Kali"**.
- The colour is usually **jet black** with white markings on tail and face and extremities sometimes found.
- **Tightly curled spiral and short horn** is an important characteristic of this breed.
- The breed is used for **grading-up of non-descript buffaloes** almost in all parts of the world and has even formed an important place in the livestock industry of many developing countries like **Bulgaria, Philippines, Malaysia, Vietnam, Brazil and Sri Lanka**. The breed has a massive body, long head and neck, short and tightly coiled horn, well developed udder and broad hips.
- Most efficient milk and butter fat producers in India.
- Bulls weigh around 550 kg (1,210 lb) and cows around 450 kg (990 lb).
- **Butter fat content is 6.9 to 8.3%. Average lactation yield is varying from 1500 to 2500 kgs per lactation.**
- Also used for the grading up of inferior local buffaloes in India.
- More recently, a Murrah buffalo named "Reshma" achieved an impressive feat by yielding 33.8 liters of milk in 24 hours, as certified by the National Dairy Development Board (NDDB).
- Total population is 4,70,65,448 on 20th Livestock Census, 2019, and it shares 42.8% among all Indian breeds of buffalo.

Fig. 1: Murrah breed: buffalo cow and bull

2. Nili-Ravi

- The home tract of Nili Ravi buffaloes is the **belt between the Sutlej and Ravi rivers of the undivided Punjab Province**.

- Actually, Nili and Ravi were two different breeds long before, but due to the passage of time and with intensive crossbreeding, the two breeds converted into a single breed named Nili Ravi.
- **Nili-Ravi is a breed of domestic water buffalo of Punjab**.
- It is distributed principally in Pakistan and India, concentrated in the Punjab region.
- Nili Ravi buffaloes are found in almost all the districts, with major concentration in Amritsar, Gurdaspur and Ferozepur districts of Indian Punjab and in Lahore, Sheikhupura, Faizabad, Okora, Sahiwal, Multan, Bohawalpur and Bahwalnagar districts of Pakistan Punjab.
- The peculiarity of the breed is the wall eyes.
- Head is small, elongated, bulging at top and depressed between eyes.
- Horns are very small and tightly coiled.
- Bullocks are good for heavy trotting work.
- Nili Ravi is also known as Panch Kalyani.
- Nili-Ravi have **walled eyes and white markings on forehead**, face, muzzle, legs and tail. The most desired character of females is the possession of these white markings known as "**Panch Kalyani**".
- The name Nili is supposed to have been derived from the **blue water of river Sutlej**.
- It is similar to the Murrah breed of buffalo, and is reared mainly for dairy use. The average milk yield is approximately **2000 kg per year; the highest record yield is 6535 kg in a lactation of 378 days.**

Fig. 2: Nili-Ravi breed: buffalo cow and bull

- Total population is 2,36,249 on 20th Livestock Census, 2019, and it shares 0.2% among all Indian breeds of buffalo.

3. Banni/Kundi

- Banni buffaloes are also known as "**Kutchi" or "Kundi"**. The breeding tract includes the **Banni area of Kachchh district of Gujarat**.
- The breed is maintained mostly **by Maldharis** under locally adapted typical extensive production system in its breeding tracts.
- Banni buffaloes are trained to **graze on Banni grassland during night and brought to the villages in the morning for milking. These traditional systems of buffalo rearing have been adapted to avoid the heat stress and high temperature during the day time.**

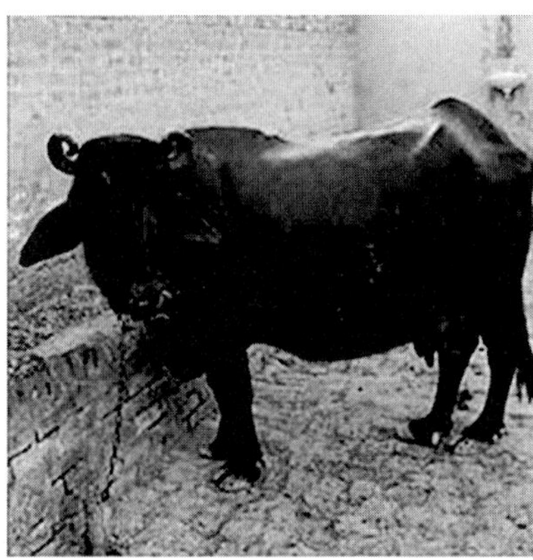

Fig. 3: Banni breed: buffalo cow and bull

- Banni buffalo have unique qualities of adaptation such as the ability to **survive water scarcity conditions**, to cover long distances during periods of drought and disease resistance.
- Banni buffaloes are mostly black in colour. Copper coloured animals are also seen in the herd.
- Forehead is elongated and straight with no slope towards the horn base.
- **Horns are tightly coiled vertically with single to double coiling.**
- The body is medium to large, compact and generally covered with hairs. Dewlap is absent and naval flap is medium. Udder is well developed, round in shape and squarely placed.

- The hind and fore quarters are uniformly well developed, whereas typically the whole udder looks like four equal divisions with teats well attached to each quarter.
- Majority of animals have conical teats with round and pointed tips. Banni buffaloes yield on an average 2857.2 kg milk in a lactation with fat % of 6.65. The lactation yield ranges from 1095 to 6054 kg.
- Total population is 7,78,466 on 20th Livestock Census, 2019, and it shares 0.7% among all Indian breeds of buffalo.

4. Surti

- Also known as **Deccani, Gujarati, Talabda, Charator and Nadiadi**.
- The breeding tract includes **Vadodara, Bharuch, Kheda and Surat** districts of Gujarat. The breed is named after its place of origin.
- Coat colour varies from rusty **brown to silver-grey to black. Skin is black or brown in colour.**
- The **horns are sickle shaped, moderately long and flat are directed downward and backward, and then turn upward at the tip to form a hook.**
- The peculiarity of the breed is **two white collars, one round the jaw and the other at the brisket region**
- The milk yield ranges from **1600 to 1800 kg per lactation**. The lactation period typically ranges from 270 to 300 days.

Fig. 4: Surti breed: buffalo cow and bull

- This breed has very high **fat percentage in milk (8-12 %), which is the** peculiarity of this breed**.** The Surti buffalo is lighter in body weight, as compared to heavy breeds, consumes less feed, thrives well both on stover and on limited or no green fodder, and produces milk with high fat and SNF content. This is popular with landless, small and marginal farmers.

- Total population is 24,52,362 on 20th Livestock Census, 2019, and it shares 2.2% among all Indian breeds of buffalo.

5. Jaffarabadi

- The breeding tract of this breed is **Gir forests, Kutch and Jamnagar districts of Gujarat**.
- **Jaffarabadi is one of the heaviest buffalo breeds** and is a native of Saurashtra region of Gujarat around Gir forest. It is also known as Bhavanagri, Gir or Jaffari.
- It is named after the town of Jaffarabad of Gujarat state. The breeding tract includes Amreli, Bhavnagar, Jamnagar, Junagadh, Porbandar and Rajkot districts of Gujarat state.
- The animals weigh up to 700 kg (average) and female animals may weigh 620 kgs (average). The animals have a big dome shaped forehead with flat, thick, downwardly curved horns.
- The **horns are heavy, inclined to droop at each side of the neck and then turning up at point (drooping horns),** forming a unique ring-like structure around the head.
- It makes eyes look small, termed as "**study eye**", especially in males. This breed is known for its ability to fight lions in Gir forest.
- The animals are generally black but some animals having white or grey tail switches are also seen.
- The udder is well developed with funnel shaped teats.
- They are very good milkers. **Average milk yield of the animals is 2239 kg per lactation with 7.7% fat. The lactation yield ranged from 2150 to 2340 kg.**
- The bullocks are heavy and used for ploughing and carting.
- These animals are mostly maintained by traditional breeders called **Maldharis**, who are nomads.

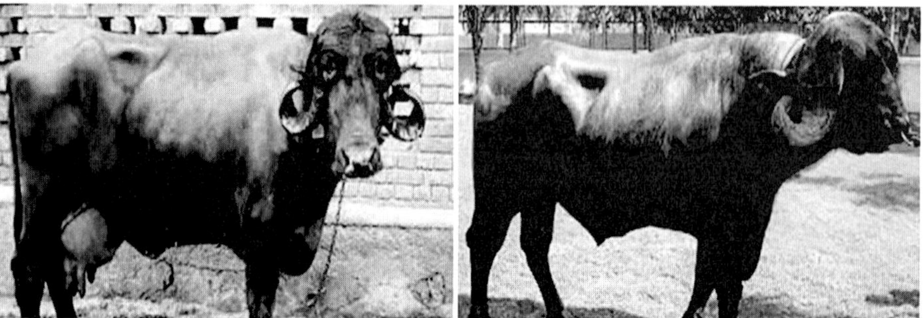

Fig. 5: Jaffarabadi breed: buffalo cow and bull

- Total population is 21,39,127 on 20th Livestock Census, 2019, and it shares 1.9 % among all Indian breeds of buffalo.

6. Mehsana

- The breed is named on the place of its origin in Mehsana district of Gujarat and also known as "Mahesani" or "Mehsani".
- Mehsana is a dairy breed of buffalo found in **Mehsana, Sabarkanda, Banaskanta, Ahmedabad and Gandhinagar districts in Gujarat and adjoining Maharashtra state.**
- The breed evolved out of **crossbreeding between the Surti and the Murrah.**
- Body is longer than Murrah but limbs are lighter.
- Bullocks are good for heavy work.
- Horns are generally sickle shaped and **less curved than Murrah buffaloes and curve more upward than Surti buffaloes.**
- Animals are mostly black in colour; a few animals are black brown or brown.
- Eyes are very prominent, black and bright bulging from their sockets with folds of skin on upper lids.
- Milk yield of the breed ranges between 598 to 3597 kg per lactation with 5.2 to 9.5% fat. The average milk yield is 1988 kg with an average milk fat is 6.83 %.

Fig. 6: Mehsana breed: buffalo cow and bull

- Total population is 43,78,788 on 20th Livestock Census, 2019, and it share 4.0 % among all Indian breeds of buffalo next after Murrah.

7. Bhadawari

- Home tract of this breed is **Agra and Etawah district of Uttar Pradesh and Gwalior district of Madhya Pradesh.**
- Also known as **"Etawah"**, Bhadawari is a dual type buffalo breed of central and northern India.
- Medium sized buffalo.
- The body is usually light or **copper coloured is a peculiarity of this breed**. Eyelids are generally copper or light brown.
- **Two white lines 'Chevron' are present at the lower side of the neck similar to that of Surti buffaloes**, called as Kanthy are present on the lower side of the neck.
- The average milk yield is 800 to 1000 kg per lactation.
- The bullocks are good draught animals with high heat tolerance.
- **This breed is an efficient converter of coarse feed into butterfat and is known for its high butter fat content.**
- On average, a Bhadawari buffalo yields approximately 4 to 5 liters of milk per day.
- **The fat percentage in their milk ranged from 8.5% to 14%, which is highest among all the buffalo breeds.**
- **The total lactation yield is lower, but the fat percentage is remarkable.**

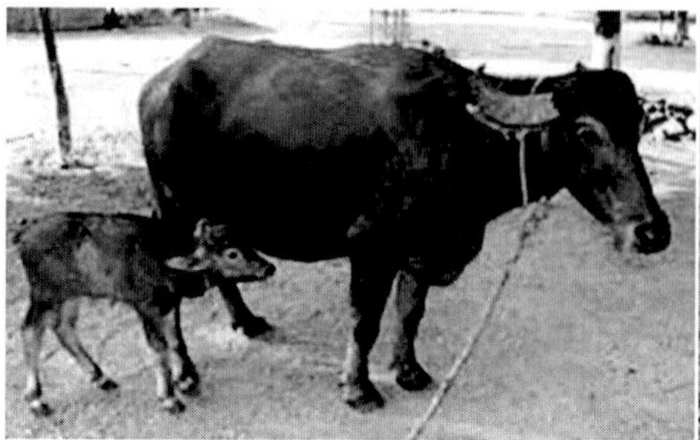

Fig. 7: Bhadawari breed: buffalo cow and bull

- Total population is 19,81,852 on 20th Livestock Census, 2019, and it shares 1.8 % among all Indian breeds of buffalo.

8. Nagpuri

- The breed has many synonyms such as "Berari", "Gaorani", Ellichpuri "Puranthadi", "Varhadi", "Gaolavi", "Arvi", "Gaolaogan", "Gangauri", "Shahi" and "Chanda".
- The breeding tract includes Akola, Amravati, Yavatmal, Wardha and Nagpur districts of Maharashtra.
- The breeding tract of this breed is **Nagpur, Akola and Amrawati districts** of Maharashtra.
- These are **black coloured animals with white patches on face, legs and tail**.
- However, "Puranthadi" strain is slightly brown in colour.
- **Horns are long flat, wide and thick at the base carried backwards in each side of the neck nearly up to the shoulders resembling a pair of swords. Length (cm) Male: 54-55 and Female: 61-62**
- The bullocks can be used for heavy work. The animals of this breed are very well-adapted to the harsh climate of Vidarbha region.
- Buffaloes and heifers in this area are reared mainly for fat production. Average milk yield per lactation is **1039 kg ranging from 760-1500 kg with average milk fat of 8.25% ranged from 7.0-8.8%.**
- Total population is 1,56,247 on 20th Livestock Census, 2019, and it shares 0.1 % among all Indian breeds of buffalo.

Fig. 8: Nagpuri breed: buffalo cow and bull

9. Toda

- This buffalo is named after an **ancient tribe, Toda of Nilgiris Hills** of south India. It is a semi-wild breed of buffalo.
- The predominate coat colours are fawn and ash-grey.
- The calves are generally fawn in colour at birth and at around 2 months of age, the fawn colour changes to ash grey. Thick hair coat is found all over the body.
- They are gregarious in nature.
- The body is long and deep. The chest is also deep. The legs are short and strong.

Fig. 9: Toda breed: buffalo cow and bull

- The **horns are set wide apart curving inward, outward and forward forming a characteristic crescent shape or semicircle.**
- **A narrow band of dense hair** covering the top line from the crest of neck to the point of origin of tail, two chevron markings, one just around the jowl and the other anterior to the brisket are also found.

- The average milk yield is 500 kg per lactation with high fat content of 8%.
- Total population is 19,868 on 20th Livestock Census, 2019. Important religious ceremonies related with the Toda buffaloes are:
- **Naming of buffaloes:** Each and every buffalo cow is named separately, soon after her first calf is born, which is followed as a routine ritual practice.
- **Exchange of buffaloes:** The buffalo cow along with its calf is exchanged during the marriage ceremony as dowry system.
- **Ceremonial slaughter:** Buffalo cows are slaughtered for ceremonial purposes to propitiate the dead relatives of the clan.
- **Migration:** It was customary that some buffaloes would migrate from one place to another in search of pasture during dry season that lasts from January to March.

10. Pandharpuri

- Pandharpuri is native breed of Maharashtra. They are named after the name of the geographical area i.e. Pandharpur block in Solapur district of Maharashtra.
- The Pandharpuri buffaloes are known to have been reared for more than 150 years in the breeding tract. **The local "Gawli" community** reared these buffaloes for milk production. These buffaloes had royal patronage from Kolhapur for supply of fresh milk to the wrestlers of this area.
- The breeding tract includes Solapur, Sangli and Kolhapur districts of Maharashtra.
- These buffaloes are concentrated in Pandharpur, North Solapur, South Solapur, Barshi, Akkalkot, Sangola and Mangalvedha tehsils of Solapur district; Miraj, Walwa, Jathand Tasgaon tehsils of Sangli district; and Karveer, Shirol, Panhala, Radhanagri, Hatkanangale and Gadhinglaj tehsils of Kolhapur district.
- **The breed is famous for its better reproductive ability, producing a calf every 12-13 months.** Under average management conditions and hot-dry climate, these buffaloes yield 6-7 litters of milk per day.
- However, under good management they are reported to yield up to 15 litters of milk in a day.

- Horns are very long and extend beyond shoulder blade, sometimes up to pin bones. These are of three types i.e. 1. Bharkand, curving back ward and usually twisted. 2. Toki, curving backward, upward and usually twisted outward. 3. Meti, flat running down.
- **They have horns are very long, running backwards, upwards and twisted outward and touching almost backbone. Horns are very long and extend beyond shoulder blade, sometimes up to pin bones.** Four types of horn orientation described which are locally called as Toki (52.05%), Bharkand (34.24%), Meti (10.81%) and Ekshing meti (2.09%). In majority of Pandharpuri buffaloes (49.92%), horn tip directed upward while in 28.23% buffaloes they are lateral.
- The Nasal bone is very prominent, long and straight.
- The buffaloes produce on an average 1790 kg of milk per lactation with 8% fat.

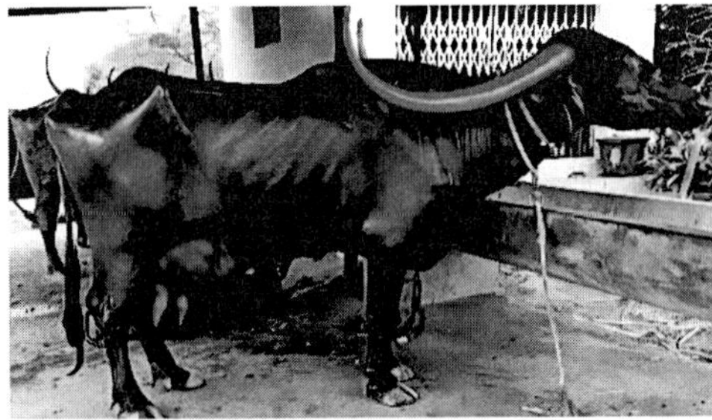

Fig. 10: Pandharpuri breed: buffalo cow and bull

- Total population is 5,14,967 according to the 20th Livestock Census, 2019, and it share 0.5 % among all Indian breeds of buffalo.

11. Jerangi

- Widely distributed in India in the **Jerangi hills of Orissa,** northern parts of Vishakhapatnam and west of Ganjam in Andhra Pradesh.
- These **small or dwarf buffaloes** are bred in the hills of Orissa.
- Its height does not exceed four feet. Horns are conical and small and run backwards; body colour is black.
- The buffaloes have a short face and small barrel. Skin is thinner and black-coloured.
- The animals are useful for ploughing water-logged paddy fields, with moderate draught capacity.

Fig. 11: Jerangi breed: buffalo bull

12. Gojri

- Buffaloes are reared in semi-migratory/pastoral management system by **Gujjar community in Pathankot, Gurdaspur, Hoshiarpur, Rupnagar and SAS Nagar (Mohali) districts of Punjab and Kangra and Chamba districts of Himachal Pradesh.**
- These buffaloes have proportionated and **medium built body and are mostly brown or black in colour**.
- Horns are **medium sized; mostly curved to form a big loop**. These buffaloes are well adapted to foot hills.
- Animals can travel long distances (seasonal migration) and can climb easily on hill tops for grazing. Used for both milk and draught power (ploughing and other agricultural operations).
- Daily milk yield ranges from 3 to 8 kg and lactation milk yield from 800 to 1200 kg. Population size is approximately 50,000.

Fig. 12: Gojri breed: buffalo cow and bull

13. Dharwadi

- **Dharwadi buffaloes are distributed in Bagalkot, Belgaum, Dahrwad, Gadg, Bellari, Bidar, Vijayapura, Chitradurga, Kalaburgi, Haveri, Kopal, Raichur and Yadgit districts of Karnataka.**
- Dharwadi is a medium sized buffalo. Coat colour is black.
- Head is straight. Horns are semi-circular and almost touching to wither.
- Ears are erect. Udder is medium in size; teats are cylindrical in shape.
- It is reared mainly for milk purpose. Average lactation milk yield is 972 liters. Daily milk yield ranges from 1.5 to 8.7 liters. Average percentage of milk fat is 6.9. The milk is used for preparation of famous Dharwad Peda with GI tag. The animals are well suited for low rainfall areas.

Fig. 13: Dharwadi breed: buffalo cow and bull

14. Manda

- Manda buffalo is distributed in **Koraput, Malkangiri and Nawarangapur districts of Odisha.**
- It is a sturdy buffalo, well adapted to hill ranges of Eastern Ghats and plateau of Koraput region of Odisha. Body colour is mostly ash grey and grey with copper-coloured hairs.
- Lower part of leg is lighter. Horns are broad, emerging slight laterally, extending backward and inward and making half circle. It is reared for draught, milk and manure.
- Both males and females are used for agriculture operations.
- **Daily milk production ranged between 1.2 to 3.7 kg. Average fat percentage in milk is 8.4.**
- Manda buffaloes are reared mostly under extensive system.

Fig. 14: Manda breed: buffalo cow and bull

15. Purnathadi

- Purnathadi buffalo name has been derived from the name of **local river Purna,** which originates in **Satpura hills and passes through Akola and Amaravati districts of Vidarbha region of Maharashtra.**
- These animals are medium in size, **whitish to light brown, while the newborn calves generally have complete whitish coat, which changes to brown as age grows.**
- Patches with white hairs are present on forehead. The lower extremities of all four legs and tail switch are white in most of the buffaloes.
- Horns are **long and tapering, may go up to the shoulder and turned upward in orientation at the end like hook**. The daily milk yield, lactation milk yield and fat percentage ranged from **1.1-5.5 kg, 353-1533 kg and 6.5-11.5, respectively**.

Fig. 15: Purnathadi breed: buffalo cow and bull

16. Bargur

- The Bargur breed of buffalo is also known as "**Malai Erumai or Malai Emmai**". Malai means hills; Erumai/Emmai means buffalo. Breeding tract is Erode district of Tamil Nadu.
- This breed is mainly used for and milk and manure.

- These buffaloes are found in the **Bargur hills in Tamil Nadu**. Coat colours vary from black to light brown or brownish-black.
- **Muzzle and eyelids are black, tail switch is brownish black and hooves are grey** in colour. Horns are Curved backward and inward.
- **Greyish white stockings from carpal/tarsal joint to fetlock are present predominantly in females.**
- **These buffaloes are maintained under the extensive system and are reared for manure, milk, and meat** (male calves are sold for carabeef).
- The animals are adapted to **graze in the hilly terrain** due to its small size (about 102 cm in height).
- The **milk yield of the animal's ranged from 1.5 to 2.0 liters per day** and of mainly used for household consumption. The average **milk fat of 8.6 %.**

Fig. 16. Bargur breed: buffalo cow and bull

17. Chhattisgarhi

- These **buffaloes are distributed throughout the Chhattisgarh state**. Coat **colour is black**. Animals are medium built with proportionate bodies.
- **Horns are medium to large in size and directed laterally backwards and then upwards with pointing tips**. These buffaloes are reared under an extensive system for providing draught power, milk and meat.
- **Males have excellent ploughing ability**, and **preferred over cow bullocks** specifically in rice fields. Milk yield ranges from 3 to 6 kg/day.
- Also known as "Desi" and mainly **used for draught, meat and milk**. Breeding tract includes Raipur, Bilaspur, Sarguja, Kanker, Mahasumund,

Dhamtari, Kavardha, Korba/Kovamha, Jashpur, Balrampur, Bemetara, Mungeli and Baloda Bazar districts of Chattisgarh.

- Common **breeding tract is north hilly, central plains and Bastar plateau region of Chhattisgarh.**
- **Chhattisgarhi buffalo bullocks are preferred over cow bullocks for ploughing the rice fields (especially during monsoon).**
- Colour of the buffalo is mostly **black but sometimes grey colour** also found. Medium sized buffaloes have straight heads, large and heavy horns which are curved laterally backwards and then going up with tip pointing upwards.
- These **buffalo are slow maturing animals and average milkers**. '**Peda**' made from the milk of these buffaloes is a famous milk product.
- The **average lactation yield of this buffalo is 1180 kg** (ranges from 1100 to 1300 kg) with an **average milk fat of 9.49 %** (ranges from 5.5 to 13.9 %).

Fig. 17: Chhattisgarhi breed: buffalo cow and bull

18. Chilika

- The breed got its name from the name of its native tract, which is **surrounding the Chilika lake in the State of Odisha.**
- The breeding tract of the breed includes **Cuttack, Ganjam, Puri and Khurda districts of Odisha**. The area mostly comprises saline zones.
- These buffaloes feed on submerged weeds and aquatic vegetation in salty waters of Chilika lake. Chilika buffalo is also known as "Deshi".
- Major utility of the breed is milk, manure and draught.
- **The colour of Chilika buffalo is brownish black or black.**

- The buffalo is **medium-sized with a compact, strong legs and small udder**. Horns are black, straight, curved upward and inward.
- Average lactation **milk yield is 500 kg** with an average **milk fat of 8.7%.** The lactation milk yield ranges from 440 to 514 kg and milk fat ranged from **8.5 to 8.8%.**
- Total population is 13,658 on 20th Livestock Census, 2019.

Fig. 18: Chilika breed: buffalo cow and bull

19. Kalahandi

- Named after the place of origin (**Kalahandi in the State of Odisha**), Kalahandi is a buffalo breed reared for milk, draught and manure.
- These buffaloes are seen in whole Gajapati district and a part of Ganjam and Rayagada district in Orissa, besides adjoining hilly regions of Andhra Pradesh.
- **Paralakhemundi** buffaloes are also known as Kalahandi buffaloes.
- It is also known as "**Deshi**". The breeding tract includes **Kalahandi and Rayagada districts of Odisha.**
- This breed is known as **Kalahandi in Orissa and Peddakimedi in Andhra Pradesh.** Reaction zone of **Paralakhemundi** buffaloes is quite distinct with strangers.
- Coat colour ranges from blackish grey to grey. **Horns go horizontally backward, upward and inward to make a half-circle appearance.**
- The muzzle, eyelids, tail and hooves are black in colour. Head is convex, hump is small and the udder is round and medium in size.

- Tail extends below the hock with coarse hairs on the switch. The animals are well-adapted to a low input extensive management system and **graze in forest, hillocks, road-side vegetation or harvested fields throughout the day**.
- Buffaloes are milked once in the morning before grazing. **Lactation milk yield is moderate, ranging from 680 to 912 kg per lactation. Milk fat percentage ranged from 7.8 to 8.2 %.**
- Females drop their first calf at about 4 years of age with lifetime calvings of 7 to 8 times with an average calving interval of 18 months.
- Paralakhemundi buffaloes are moderate milk yielders having average daily milk yield of 2.6 litre and average lactation yield of 737 litre in 285 lactation days.
- These buffaloes are known for their working ability and disease resistance in the native tract.
- Total population is 25,164 on 20th Livestock Census, 2019.

Fig. 19: Kalahandi breed: buffalo cow and bull

20. Luit (Swamp)

- They are also known as "**Assamese Swamp**". Breeding tract includes **Jorhat, Sibsagar, Lakimpur, Dibrugarh, Tinsukia, Dhemaji, Golaghat, Majuli and Biswanath districts of Assam.**
- The Luit (Swamp) buffaloes are mostly **distributed in the upper Brahmaputra valley of Assam** covering nine districts of upper Assam.
- These are swamp buffaloes having 48 diploid number of chromosome (2N) and distributed mostly in upper Brahmaputra valley of Assam.

- These buffaloes are also found in some areas of Manipur, Mizoram and Nagaland bordering Assam.
- People of Assam have been traditionally rearing swamp buffaloes in the **embankment and small islands of the mighty river Brahmaputra (Luit),** hence, these buffaloes are named Luit.
- Mainly used for milk and draught. Colour of buffalo is mainly black. These swamp buffaloes are medium-sized strongly built animals with prominent wither and short tail.
- The animals are characterized by **broad and concave forehead with prominent eyes and wide muzzle and distinct semi-circular horns in both sexes.**
- They possess **light white stockings up to the knee in both fore and hind legs. Horns are curved laterally backward and then upward forming a semi-circle**.
- The **average lactation yield of Luit buffalo is 449 kg with an average milk fat of 8.68 %.**
- As a large proportion of these buffaloes are reared on river islands, river banks and forest land in large groups, locally called Khuti which means nomadic.

Fig. 20. Luit breed: buffalo cow and bull

21. Marathwadi

- They are also known as "**Dudhana Thadi**". The breeding tracts include **Parbhani, Nanded, Beed, Jalna and Latur districts of Maharashtra**.
- The breed is maintained for milk as well as draught/transport purposes. The animals are greyish black to jet black in colour and white markings are sometimes present on the forehead and lower parts of the limbs.

- Horns are medium in length (average 43 cm) and parallel to the neck, reaching up to shoulder but never beyond shoulder blade.
- Distinguishing feature from Pandharpuri breed is the **length of the horns, which reach only up to shoulder**, **unlike in Pandharpuri breed wherein they may even reach up to pin bones sometimes.**
- The animals are reared in a semi-intensive management system in a mixed herd comprising cattle and buffaloes.
- They are usually fed sorghum, paddy straw, sugarcane leaves/tops and grasses.
- Milking animals are offered concentrate. **Average lactation milk yield is 1118 kg and average milk fat is 8.8% ranging from 6.25-10.50%.**
- Total population is 2.42.650 on 20th Livestock Census, 2019 and it share 0.2 % among all Indian breeds of buffalo.

Fig. 21: Marathwadi breed: buffalo cow and bull

22. Godavari

- This is a lesser known breed of buffalo, having their breeding tract in east and west Godavari districts of Andhra Pradesh. This breed has their **origin from interbreed crossing, followed by grading up of local non-descript buffaloes with Murrah breed**.
- The animals of this breed are also found in areas of Tanuku, Bhimavaram, Narasapur, Ramachandrapuram, Kothapeta, Alamurualuqa and part of Tadepanigudem and Kovvuru, Krishna deltaic areas of Gudlavalluru.
- The animals are medium-statured with a compact body.
- Colour is predominantly black with a spare hair coat of coarse brown hairs.

- The horns are short, flat, curved, slightly downwards, backwards and then forward with a loose ring at the tip.
- Udder is medium in size, bowl-shaped and well-paced medium-sized teats. Milk yield is around 2050 kg in a lactation of 305 days.

Fig. 22: Godavari breed: buffalo

23. Tarai

Origin and breeding tract: Tarai buffalo breed is the mainly breed of Uttarakhand. This breed is found mainly in Ram Nagar, Sitarganj, Khatima, Nanakmatta areas of Kumaon of Uttarakhand. This breed is also found in Philibhit and Bareilly districts of Uttar Pradesh.

Population: As per 18th livestock Census 2007, the total number of animals of this breed was recorded as 27,757 and total population of buffaloes in the state was recorded as 12-19 lacs. Tarai breed of buffalo is less known breed being maintained by farmers in Tarai part of Uttarakhand and U.P. State.

- Some of the important characteristics is described as follows:

Body color	Grayish brown to black coat colour on black skin. Eyelids, hooves, nose ridge, face and pastern joint are mostly black in 89 to 95% cases.
Horns	Orientation of horns is mostly backward (76%) and black in color. Shape of horn is slightly curved to sickle shape (76%) measuring 44 cm with tip projecting mostly upward (72%). Horns are smaller in females (32 cm) than males (44 cm.).
Tail	Slightly lower than hock joint measuring 89 to 99 cm. Switch of the tail is mostly white.
Body	Medium, compact with small, straight and shining hair.Navel is tight and sheath is non pendulous.
Head	Short in length, convex (60%) or flat (35%) in shape

Neck	Strong neck of 63 to 66 cm long in female and 70 to 71 cm long in males.
Ear	Orientation of ear is backward and comparatively small in size.
Udder	Shape of udder is mostly round (56%) and rarely pendulous. Fore udder is either flat (31%) or projected (45%) when filled with milk. Rear udder is small.
Teats	46% of teats are small with pointed tips.

- Heart girth of adult animals is 201 cm, while paunch girth is 227 cm, height of the animals is 156 cm. Navel is tight and sheath is non pendulous. Temperament of female is docile, however, that of male is aggressive as usual.
- Fat per cent of milk was recorded from records of the primary milk collection society where farmer was supplying the milk.

(a) Monthly milk yield: Milk yield on day of recording multiplied by 30 was taken as monthly milk yield. Average milk yield during first month of first lactation was 127.07±4.18kg. while during second lactation it was 125.86±3.39 kg. during second month. Milk yield progressively decline from third month to 12th month of lactation.

(b) Lactation milk yield: Average first lactation milk yield recorded as 1030.04±26.78 kg while that in second lactation it was 1080.09±28.58 kg. Thus, there was an increase in milk production by 4.86% during second lactation over first lactation. Over all average milk yield in both lactations was 1054.08±1957 kg.

(c) Fat Percent: Fat per cent recorded during first lactation was 6.35±0.11 per cent while in second lactation it was 6.67±0.14 with over all fat per cent of 6.58±0.09%.

(d) SNF content: Average SNF calculated during first lactation was 8.56±0.08 per cent while 8.59±0.02 during second lactation. The overall SNF content was 8.57±0.05 based on 47 samples.

(e) Lactation length: Seventy nine per cent of buffaloes completed lactation up to 9 months while only 12.6% of buffaloes remained in lactating upto one year period. The overall lactation length was recorded as 291.19±3.63 days.

Reproduction Performance

(a) Service Period: Based on sample of 201 buffaloes, service period in Tarai buffalo was recorded 197.07±6.59 days with 47.45% coefficient of variation.

(b) Dry Period: Dry period was observed as 186.92±16.77 days with 46.62 per cent coefficient of variation.

(c) Calving interval: Average calving interval noted under the recording period was 470.62±18.07 days with 19.95 per cent coefficient of variation.

In the **breeding tract of Tarai buffaloes**, most of the farmers are keeping one or two buffaloes and land holding is small. So it is not possible for each farmers to maintain a breeding bull. Thus, in the area, **a few Tarai bulls are maintained by the farmers have to be used for breeding**.

Conservation of breed

(i) Ten thousand Five hundred Ninety (10,590) doses of deep freeze semen of Tarai buffalo were prepared and supplied to NBAGR, Karnal.

(ii) Twenty Eight (28) Tarai buffalo bulls maintained under the project were transferred to U.S. Nagar Dugdha Utpadak Sahakari Sangh Ltd., Rudrapur for distribution to different milking societies for use in breeding and improvement of Tarai buffaloes.

Fig. 23: Tarai breed: buffalo cow and bull

3

Sheep Breeds of India

According to the karyotyping the domestic sheep (*Ovis aries*) has 54 chromosomes. The following is a brief summary of their chromosome composition:

Autosomes: There are one pair of sex chromosomes, 52 autosomes and three pairs of these autosomes are submetacentric (having a centromere slightly off-centre), while the rest are acrocentric (with the centromere near one end).

Sex Chromosomes: The X chromosome is acrocentric and the Y chromosome varies from submetacentric to metacentric.

Important features

- According to the Food and Agriculture Organization Corporate Statistical (FAOSTAT) Database (2019) and database of the United Nations Food and Agriculture Organization, the top three countries by the number of heads of sheep were: China (163.48 million heads), India (74.26 million) and Australia (65.75 million).
- India has second largest sheep population in the World.
- Sheep milk is an important product in the Near East and North Africa (9% of total milk production) and sub-Saharan Africa (4%).
- The countries with the most dairy sheep are China, Sudan, Turkey and Algeria (FAO).
- Major sheep milk producers are China, Turkey and Greece.
- More than half of the world's sheep population is found in developing countries. Sheep are more prevalent than goats in cold climatic conditions. The major produce from Sheep production are milk, meat, skin, fibre and manure. However, most of the small-scale producers in developing countries raise sheep for meat or sale.

- Maximum sheep milk is produced in the Mediterranean region, and majority of dairy sheep breeds are found in this region and the Near East. The milk yield and lactation length of dairy sheep are not comparable to those of dairy cattle or dairy goats, but sheep milk production can be improved by milking stimulation (e.g., milking several times a day).
- Genetic selection of dairy sheep has not resulted in significant improvements in milk yield and lactation length. Dairy sheep breeds include Awassi, East Friesian and Lacaune (FAO).

The scientific classification of sheep are

Kingdom: Animalia

Phylum: Chordata

Class: Mammalia

Order: Artiodactyla

Suborder: Ruminantia

Family: *Bovidae*

Subfamily: *Caprinae*

Genus: *Capra* (goat) or *Ovis* (sheep)

Species: *Capra aegagrus hircus* (goat) or *Ovis aries* (sheep)

India has a vast genetic resource and currently has 45 registered breeds of sheep (NBAGR). All these breeds are classified according to geographical locations.

(i) Northern Temperate Region

- This region comprises Jammu and Kashmir, Himachal Pradesh, Uttarakhand and hilly parts of Uttar Pradesh. Fine and coarse wool breeds are found in this zone. The important breeds of this region are as follows:

Sl. No.	Breeds	State
1	Gaddi (CW)	Himachal Pradesh
2	Rampur Bushair (CW)	Himachal Pradesh
3	Gurez (CW)	Jammu and Kashmir
4	Karnah (AW)	Jammu and Kashmir
5	Poonchi (CW)	Jammu and Kashmir
6	Bhakarwal (CW)	Jammu and Kashmir
7	Changthangi (CW)	Jammu and Kashmir

CW=Carpet wool, AW=Apparel wool

(ii) North Western region

This region comprises the states of Punjab, Haryana, Rajasthan and Gujarat and the plains of Uttar Pradesh, Uttarakhand and Madhya Pradesh. Coarse and carpet wool sheep breeds are found in this zone as follows:

Sl. No.	Breeds	State
1	Patanwadi (CW)	Gujarat
2	Panchali (MCW)	Gujarat
3	Kajali (MCW)	Punjab
4	Chokla (CW)	Rajasthan
5	Jaisalmeri (MCW)	Rajasthan
6	Magra (CW)	Rajasthan
7	Malpura (MCW)	Rajasthan
8	Marwari (MCW)	Rajasthan
9	Nali (CW)	Rajasthan
10	Pugal (MCW)	Rajasthan
11	Sonadi (MCW)	Rajasthan
12	Jalauni (MCW)	Uttar Pradesh and Madhya Pradesh
13	Muzaffarnagri (MCW)	Uttar Pradesh and Uttarakhand

MCW=Meat and Carpet Wool

(iii) Southern peninsular region

This region (semi-arid in central peninsular and hot humid region along the coast) comprises of Maharashtra, Andhra Pradesh, Karnataka, Tamil Nadu and Kerala. The sheep breeds found in this region are as follows:

Sl. No.	Breeds	State
1	Nellore (M)	Andhra Pradesh
2	Deccani (M)	Andhra Pradesh and Maharastra
3	Hassan (M)	Karnatak
4	Bellary (MCW)	Karnatak
5	Kenguri (M)	Karnatak
6	Mandya (M)	Karnatak
7	Coimbatore (MCW)	Tamil Nadu
8	Katchaikatty-Black(M)	Tamil Nadu
9	Kilakarsal (M)	Tamil Nadu
10	Madras-Red (M)	Tamil Nadu
11	Macheri (M)	Tamil Nadu
12	Nilgiri (AW)	Tamil Nadu
13	Ramanadhapuram-white (M)	Tamil Nadu
14	Trichi black (M)	Tamil Nadu
15	Vembur (M)	Tamil Nadu
16	Chevaadu (M)	Tamil Nadu
17	Macherla (M)	Andhra Pradesh

M=Meat

(iv) Eastern Region

This region (hot and humid) includes Bihar, West Bengal, Orissa, Assam and other eastern states. The sheep in this region are primarily of meat type but Arunachal Pradesh has a small number of better-quality wool sheep. **Garole breed of Sundarban of West Bengal has high fecundity rate.** Other important breeds of this region are Balangir, Bonpala, Ganjam, Tibetan.

Sl. No.	Breeds	State
1	Tibetan (CW)	Arunachal Pradesh
2	Shahbadi (MCW)	Bihar
3	Chottnagpuri (MCW)	Jharkhand
4	Balangir (MCW)	Odisha
5	Ganjam (MCW)	Odisha
6	Kendrapada (MCW)	Odisha
7	Bonpala (MCW)	Sikkim
8	Garole (M)	West Bengal

Description of different sheep breeds

I. Northern Temperate Region: This region comprises of Jammu & Kashmir, Himachal Pradesh, Uttarakhand and hilly parts of Uttar Pradesh. The important breeds of this region are Bhakarwal, Changthangi, Gaddi, Gurez, Karnah, Poonchi and Rampur Bushair.

1. Bhakarwal

- The name of the Bakharwal breed is derived from the nomadic tribe which rears these sheep.
- The breed has no distinct home tract, and the sheep are entirely migratory. Bakharwal sheep flocks, winter in the Pir Panjal ranges of the Jammu division, and in the summer migrate to the Kashmir Valley, crossing the high mountain passes. They are medium-sized animals, with a typical Roman nose.
- The fleece which is coarse and open is generally white, although coloured fleeces are occasionally observed.
- They are a coarse carpet wool breed of sheep. All animals are spotted fawn or grey. The rams are horned and the ewes are polled.
- The ears are long and drooping, and the tail is small and thin. Adult ewes weigh between 29 and 36 kg; rams can weigh as much as 55 kg.

- Most of this breed has now been crossed with Merino for improving greasy-wool production and quality for apparel wool and only a small proportion of flocks still contain pure Bakharwal animals.
- The sheep are shorn three times a year. The total annual wool produced per animal ranges from 1 to 1.5 kg.
- Population size 1, 87, 967 as per 20th Livestock Census, 2019.

Fig. 1: Bhakarwal sheep breed: ewe and ram

2. Changthangi

- Changthangi sheep is a native breed of Changthang region of Ladakh and is locally called Changluk sheep.
- **This sheep is mainly reared by a nomadic tribe called Changpa along with Pashmina producing Changthangi goats.**
- Changpas are semi-nomadic tribal people found mainly in the Changthang in Ladakh of Jammu and Kashmir.
- The Changpas are always on the move with their animals. They mostly rear yak, Pashmina goats and Changluk sheep. Changluk sheep act as main source of income for this tribe.
- Shepherds of the Changpa tribe take the herds in the morning and roam the area far and wide, while whistling and singing, in search of good pastures and water. In the evening before the sun sets the shepherds bring the herd back home.
- The entire family is involved in taking care of the sheep and the herd. Breed Description Changthangi sheep is a medium size sheep with good fleece cover. The body of these sheep is covered with outercourse hair covering called guard hair, beneath, which is a fine fibre coat.

- Average weight of adult ram ranges from 50-70 kg while that of ewes range from 35-50 kg. Majority of the animals have white colour. Other colours include a wide range of shades like black, brown, and grey, but white colour dominant in the sheep flock.
- The Ewes are very protective and attentive mothers resulting in high lamb survival rate. Both horned and polled animals are found in Changluk sheep. The tail is medium to short in length. Population size is 71,901as per 20th Livestock Census, 2019.

Fig. 2: Chnagthangi sheep breed: ewe and ram

3. Gaddi

- **The Gaddi breed, is also known as Bhadarwah, which is native to the Kishtwar and Bhadarwah Tehsils in the Jammu region of Jammu and Kashmir.** The breed is distributed in Kishtwar and Bhadarwah Tehsils in Jammu province of Jammu & Kashmir state, **Hamirpur, Ramnagar, Udhampur and Kulu and Kangra valleys of Himachal Pradesh and Dehradun**, Nainital, Tehri Garhwal and Chamoli districts of Uttarakhand.
- These are medium sized animals, usually white, although tan, brown and black and mixtures are also found.
- Males are entirely horned but females to the extent of only 10 to 15% are horned.
- Fleece is generally white with brown coloured hair on the face. **Wool is fine and lustrous; average annual yield is 1.13 kg per sheep, clipping is done thrice in a year.**
- A part of this clip is sent to Dhariwal mills and Amritsar markets. Undercoat is used for the manufacture of high quality Kulu shawls and blankets.

Fig. 3: Gaddi sheep breed: ewe and ram

- Population size is 6,66,915 as per 20th Livestock Census (2019).

4. Gurej

- The Gurej breed is found in the **Gurez block of Bandipore district in North Kashmir**. They are found in largest number in the state.
- The skin colour is pink.
- Both sexes are polled. Ears are large and leafy; the tail is short to medium in length and thin.
- Fleece is **white coarse, dense and long stapled**. Forehead, belly and legs are covered with wool. The March and September clips are yellow but the September clip is golden yellow in colour.
- Body colour is predominantly white with brown or black colour on muzzle and around eyes. Complete coloured sheep may also be seen.
- The overall averages for CWY (%), FD (μ), SL (cm), CPI, and M (%) were 76.10±0.13, 28.31±0.17, 4.48±0.07, 5.37±0.10 and 9.29±0.39, respectively.
- Population size is 25,714 as per 20th Livestock Census (2019).

Fig. 4: Gurej sheep breed: ewe and ram

5. Karnah

- The Karnah breed is primarily found in **Karnah, a mountainous tehsil of Kupwara district in North Kashmir.**
- These are generally large animals. The rams have large curved horns and a prominent nose line.
- **Wool is generally white in colour.**
- The sheep are shorn twice in a year i.e., spring and autumn. The wool production ranged between 1 to 1.5 kg per animal per year. Staple length ranged from 12 to 15 cm and the average fibre diameter between 29 to 32μm.
- Population size is 3,121 as per 20th Livestock Census (2019).

Fig. 5: Karnah sheep breed: ewe and ram

6. Poonchi

- The Poonchi breed, native is **Poonch and Rajouri districts of the Jammu region of Jammu and Kashmir.**
- The animals are similar in appearance to **Gaddi except being lighter in weight.**
- Animals are predominantly **white in colour, including the face but spotted sheep varying from brown to light black are also seen.**
- Ears are medium long. Tail is short and thin. Legs are also short, giving a low-set conformation.
- The weight of the adult ram ranges from 35 to 40 kg and that of an ewe from 25 to 30 kg. Wool is of medium to fine quality and mostly white in colour.

- Sheep are shorn three times in a year and produce between 0.9 to 1.3 kg greasy wool per year. Fibre length ranges between 15 to 18 cm and the fibre diameter between 22 to 30 μm.
- Population size is 20,232 as per 20th Livestock Census (2019).

Fig. 6: Punchi sheep breed: ewe and ram

7. Rampur Bushair

- **The Rampur Bushair breed is distributed in Shimla, Kinnaur, Nahan, Bilaspur, Solan and Lahaul and Spiti districts of Himachal Pradesh and Dehradun, Rishikesh, Chakrata and Nainital districts of Uttarakhand.**
- These are medium-sizedanimals. The fleece colour is predominantly white, but brown, black and tan colour are also seen on the fleece in varying proportions.
- The ears are long and drooping. The face line is convex, giving a typical Roman nose.
- The males are horned but most of the females are polled.
- The **fleece is of medium quality and dense**. Legs, belly and face are devoid of wool.
- Population size is 18,239 as per 20th Livestock Census (2019).

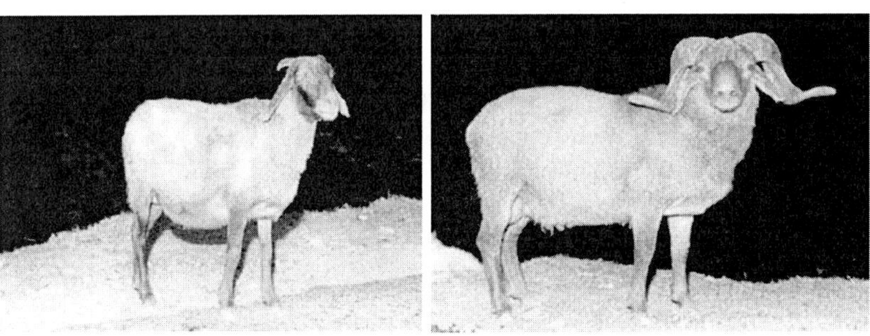

Fig. 7: Rampur Bushair sheep breed: ewe and ram

- **Average adult body weight:** Males: 34.7 kg, Females: 26.0 kg.
- **Body weight range:** Adult males: 24 to 47 kg, Adult females: 18 to 40 kg.
- **Wool yield:** Varies from 200 to 500 g per shearing; they are shorn twice a year (summer and winter).
- **Sexual maturity:** Males and females reach sexual maturity at one year of age.

II. North Western region: This region comprises the states of Punjab, Haryana, Rajasthan and Gujarat and the plains of Uttar Pradesh, Uttarakhand and Madhya Pradesh. Important breeds of sheep found in this region are Chokla, Jaisalmeri, Jalauni, Magra, Malpura, Marwari, Muzaffarnagri, Nali, and Patanwadi.

1. Chokla

- Synonyms: Chapper, Shekhawati and Raata Munda
- Origin: The name Shekhawati and Chhappar are derived from the name of its distribution area whereas Raata munda stands for dark brown coloured face.
- Chokla sheep found mostly in Shekhawati, Churu, Nagaur and Sikar.
- Chokla breed is considered to be the best wool breed in Rajasthan and India, hence **Chokla sheep is also called Merino of India**.
- Chokla is perhaps the **finest carpet-wool breed**, although most Chokla wool is now being diverted to the worsted sector because of a lack of fine apparel-wool in the country.
- Cross-breeding of the Chokla with exotic fine wool breeds (Merino and Rambouillet) has been promoted to improve apparel-wool production and quality, and this has led to a decline in numbers of this breed.
- They are light to medium-sized animals. **The face, generally devoid of wool, is reddish brown or dark brown, and the colour may extend up to the middle of the neck; the skin is pink.**
- The ears are small to medium in length and tubular. Both sexes are polled. The tail is thin and of medium length.

- The coat is dense and relatively fine, covering the entire body including the belly and the greater part of the legs. Age at first breeding for males is 15 months and for females is 24 months.
- Lambing percentage is about 75 and the litter size is single with rare cases of twinning. A large percentage of males not required for breeding are castrated and kept for wool production. Male lambs are sold for slaughter between the ages of 5 to 7 months.
- The annual average fleece weight is 2.74 kg with an average fibre diameter and density of 28 µm (micron) and 1,040 cm^2, respectively, and a medullation percentage of 24.
- The Chokla sheep population in India is estimated to be approximately 3,82,197 as of the 20th Livestock Census (2019).

	Adult Male	Adult Female
Average body weight (kg)	41	31
Average body length (cm)	71	63
Average height at withers (cm)	69	63
Average chest girth (cm)	82	75

Fig. 8: Chokla sheep breed: ewe and ram

2. Jaisalmeri

Fig. 9: Jaisalmeri sheep breed: ewe and ram

- The name of the Jaisalmeri breed is derived from its home tract, Jaisalmer. They are distributed **across the Jaisalmer, Barmer and Jodhpur districts of Rajasthan**.
- Jaisalmeri female pure specimens are found in **south western Jaisalmer, extending up to north western Barmer and southern and western Jodhpur**.
- They are primarily reared by the **Raikas and Sindhi Muslim communities**. In terms of body size, this is the largest breed among the eight breeds of Rajasthan, and produces good quality carpet-wool.
- They are tall, well-built animals, **with black or dark brown faces**, with the colour extending up to the neck. Other characteristics include a typical Roman nose and long drooping ears, generally with a cartilaginous appendage.
- **Both sexes are polled**. The tail is medium to long and the fleece colour is white. Age at first breeding is 15-18 months for rams and 12-15 months for ewes.
- **Lambing percentage is about 55 to 60**. Litter size is usually single. Rams are selected on the basis of fleece weight for breeding and are usually bred by flock owners; in some cases, they are exchanged with other owners.

- **The ewes are not milked. Shearing is done three times in a year.**
- The average greasy fleece weight per clip is 753 gm with an average fibre diameter of 35 μm and a medullation percentage of 40.
- According to the 20th Livestock Census the total number of Jaisalmeri sheep in the country is 6,80,173.

	Adult Male	Adult Female
Average body weight (kg)	54	34
Average body length (cm)	77	68
Average height at withers (cm)	76	68
Average chest girth (cm)	95	82

3. Jalauni

Fig. 10: Jalauni sheep breed: ewe and ram

- **Jalauni sheep are found in the Bundelkhand region of Uttar Pradesh and Madhya Pradesh states in India.** About 37% of the geographical area of this region is under cultivation and about 86% of the population, mostly directly or indirectly dependent on agricultureand reside in the villages.
- Forests, cultivable waste and barren land occupy more than 50% of the area, permanent pasture and other grazing lands about 9% and miscellaneous tree crops and grasses about 0.7%.
- The livestock census figures of 1977 and 1997 indicate an annual declining trend of 0.04 % in the sheep population in Uttar Pradesh and Madhya Pradesh states.

- Characterization and evaluation of Jalauni sheep under field conditions were undertaken to establish the norms, morphological characteristics and performance parameters.
- A total of 78 households were visited in 29 villages in five districts. Information on feeding, breeding and management practices, utility patterns etc. and production and reproductive performance was collected through personal observations and interaction with the farmers.
- Body weight and/or body measurements were recorded on 374 animals. Average adult body weights of male and female Jalauni sheep were 35.5±2.1 kg and 27.2±0.7 kg, respectively.
- **Age at first lambing was 1.5 to 2 years and the lambing interval was one year. An ewe, on an average, delivers 7–9 lambs in her lifetime.**
- Population size is 42, 931 as per 20th Livestock Census (2019).

4. Magra

- The Magra sheep breed, also known as **Bikaneri Chokhla or Chakri and formerly known as the Bikaneri, is found in the Bikaner, Nagaur, Jaisalmer and Churu districts of Rajasthan, India.** However, purebreds are only found in the eastern and southern parts of the Bikaner district.
- Synonyms: Magreti, Bikaneri Chokhla, Raata chakriya, Chakri, Boochie kan and Desi.
- **The Magra sheep is known as the only lustrous carpet wool-producing breed.** The most important strain of Magra has flocks with extremely white and lustrous fleece, which can only be found in a few villages around Bikaner.
- A breeding program aims to improve this breed through selection. The breeders, wool traders, and industrialists in Bikaner have expressed **serious concern at the rapid decline in numbers of Magra sheep.**
- Moreover, the number of purebreds is decreasing due to crossbreeding with other breeds in the area. It has been observed that there is a serious need for its conservation.
- Population size is 1,31,689 as per 20th Livestock Census (2019).

Fig. 11: Magra sheep breed: ewe and ram

5. Malpura

- Malpura sheep are found in **Jaipur, Tonk, Sawaimadhopur and adjacent areas of Ajmer, Bhilwara and Bundi districts in Rajasthan.**
- The animals are fairly well-built, with long legs and light brown faces. Ears are short and tubular, with a small cartilaginous appendage on the upper side. Tail length is medium to long and thin. Both sexes are polled.
- The **fleece is white, extremely coarse and hairy. Belly and legs are devoid of wool**. The overall least square means for first six monthly and adult annual GFY were found as 551 g and 810 g, respectively. The least square means of birth, 3-, 6- and 12-month's weight of lambs were recorded as 3.02, 15.41, 20.80 and 25.60 kg, respectively under farm conditions.
- The average fibre diameter, medullation and staple length were 41.67μm, 75.9 % and 4.9cm, respectively.
- Population size is 2,09,534 as per 20th Livestock Census (2019).

Fig. 12: Malpura sheep breed: ewe and ram

6. Marwari

- Marwari sheep are distributed in **Jodhpur, Jalore, Nagaur, Pali and Barmer districts** extending up to Ajmer and Udaipur districts of Rajasthan and the Jeoria region of Gujarat.
- The animals are **medium in size with black face, this colour extending to the lower part of neck.** Ears are extremely small and tubular. Both sexes are polled.
- Tail is short to medium and thin. The fleece is white and not very dense.
- The overall least square means for greasy fleece weights for first six monthly, adult six monthly and adult annual were 607.16, 631.25 and 1260.50g, respectively. The least square means of birth, 3-, 6-, 9- and 12-month's weight of lambs were 3.05, 14.74, 19.33, 22.85 and 25.90 kg, respectively under farm conditions.
- Average fibre diameter, medullation and staple length were 31.9μm, 50.8 % and 5.35cm, respectively.
- Population size is 28,70,057 as per 20th Livestock Census (2019). **This breed ranks third among all goat breeds of India having 4.1 % population.**

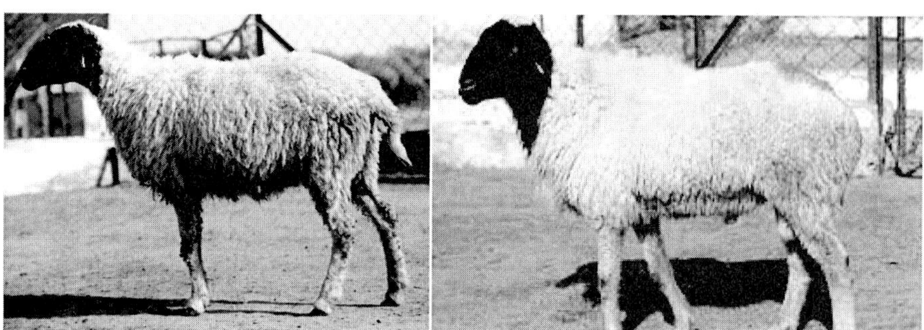

Fig. 13: Marwari sheep breed: ewe and ram

7. Muzaffarnagari

- **This breed is distributed in Muzaffarnagar, Bulandshahr, Saharanpur, Meerut, Bijnor districts of Uttar Pradesh and Dehradun district of Uttarakhand and parts of Delhi and Haryana.**
- Pure specimens are found in Muzaffarnagar district. The animals are medium to large in size.

- Face line is slightly convex. Ears and face are occasionally black. Both sexes are polled. Males occasionally show rudimentary horns.
- Ears are long and drooping. Tail is extremely long and reaches a fetlock.
- The fleece is white, coarse and open. Belly and legs are devoid of wool. The overall least square means for lamb's first and second six monthly and adult annual clips were 582.17, 538.75 and 1217.62g, respectively under farm conditions.
- The overall averages of weight at birth, 3, 6, 9 and 12 months of age were 3.63, 15.59, 23.52, 27.14 and 30.76kg, respectively. . Least square means of wool quality attributes viz. fibre diameter, hetero fibres, hairy fibres, medullation and staple length were 38.39±1.36μm, 14.35±1.37%, 46.89±3.42%, 61.03±4.33% and 5.09±0.30cm, respectively.

Fig. 14: Muzaffarnagari sheep breed: ewe and ram

- Population size 41,760 as per 20th Livestock Census, 2019.

8. Nali

- **The Nali sheep is found in Ganganagar, Churu and Jhunjhunu districts of Rajasthan, southern part of Hisar and Rohtak districts of Haryana.**
- The animals are **medium-sized. Face colour is light brown and skin colour is pink. Both sexes are polled. Ears are large and leafy.**
- Tail is short to medium and thin. **Fleece is white, coarse, dense and long-stapled. Forehead, belly and legs are covered with wool.**
- The overall means of fleece weight for first six monthly and adult annual were 1.01 and 2.84 Kg, respectively. The means of birth, 3, 6, 9 and 12-month's weights of lambs were 2.43, 10.74, 14.93, 17.13 and 19.64 kg, respectively under farm conditions.

- Average fibre diameter, medullation and staple length were 29.89μm, 41.14 % and 6.79cm, respectively.
- Population size is 2,50,343 as per 20th Livestock Census (2019).

Fig. 15: Nali sheep breed: ewe and ram

9. Patanwadi

- Patanwadi sheep is a fascinating breed found in **western India, particularly in the coastal plains of Saurashtra and Kutch regions in Gujarat**.
- **Non-Migratory Patanwadi**: These are **red-faced** animals with small bodies. They are typical Patanwadi and are primarily located in north-eastern Saurashtra.
- **Migratory Type:** This strain exhibits a **larger body and long legs**. They have a distinctive Roman nose and long tubular ears.
- **Meat Type:** The meat type Patanwadi have a big body and low stature.
- Patanwadi is also called **desi, Kutchi, Vadhiyari and Charotari**. These animals, in general, are medium to large with relatively long legs.
- They have **typical Roman nose with a brown face** which may be tan in a few cases. Ears are drooping, medium to large, tubular with a hairy tuft.
- The tail is thin and short. **Both sexes are polled**. The overall least square means for first six monthly and adult annual GFY were 566 and 1063g, respectively and least square means of birth, 3-, 6-, 9- and 12-month's weights of lambs were 3.23, 14.26, 17.67, 20.76 and 24.22 kg, respectively under farm conditions.

- Average fibre diameter and medullation were 29.11µm and 38.41%, respectively.
- Population size is 8,79,620 as per 20th Livestock Census (2019).

Fig. 16: Patanwadi sheep breed: ewe and ram

10. Panchali

- Panchali is a **dual-purpose sheep reared for milk and meat** in **Panchal area of Gujarat**.
- Synonyms: Baraiya, Dooma, Dumma, Panchali-Dumma.
- Animals are large in size, and have long legs and excellent migration ability.
- Coat colour is white. **Head or facial parts are black, blackish brown, brown and light brown in colour.**
- Ears are long and pendulous. Tail is long. Udder is well-developed.
- Milk yield ranged from 0.4 to 1.2 litre/ day. Animals attained 18 to 20kg body weight at 3 to 4 months of age. Adult weight varies from 53 to 82kg in males and 32 to 73kg in females.
- Wool is coarse and annual production is nearly one kg.

Fig. 17: Panchali sheep breed: ewe and ram

11. Kajali

- Kajali sheep, also known as Kajli, is a breed native to both India and Pakistan.
- Origin and distribution: India, **Kajali sheep are distributed in Punjab, India.**
- Pakistan: They are mainly bred in cities such as Sargodha, Gujranwala, Lahore, Faisalabad, Toba Tek Singh, and Jhang.
- Appearance: **Kajali sheep are known for their beauty and can grow significantly large and bulky at a young age.**
- Purpose: They are bred **primarily for their wool, meat, and milk.**
- Coloration: Kajali sheep have a **white body coat with a black circle around the eyes and a black tip extending to the lower one-third of the ears.**
- Livelihood: Indigenous sheep breeds like Kajali play an essential role in securing the livelihood of small and marginal farmers. They are maintained on low/zero input extensive management systems.
- Genetic Resources: Despite declines in total sheep populations, Kajali sheep contribute to the rich ovine genetic resources of both countries. Detailed characterization and documentation of lesser-known breeds like Kajali are crucial for conservation and sustainable utilization.
- **Mutton Quality: Kajali sheep are well-known for their juicy mutton quality.**

Fig. 18: Kajali sheep breed: ewe and ram

12. Pugal

- Pugal area of Bikaner district is its home tract. It is **distributed over Bikaner and Jaisalmer districts of Rajasthan.**
- **Synonyms: Rataanaa.**
- The animals are fairly well built. **Face is black with small light brown stripes on either side above the eyes**; the lower jaw is typically light brown.
- The black colour may extend to the neck. Ears are short and tubular. Both sexes are polled. Tail is short to medium and thin. The fleece is white.
- **Appearance:** Black face with a light brown colour on the lower jaw, tubular and short ears, short and thin tail and less dense white fleece.
- **Production traits:** Average milk yield: 300-500 grams per day, weight of ram: Approximately 32 kg, weight of ewe approximately 27 kg and average wool yield 1.60 kg.

Fig. 19: Pugal sheep breed: ewe and ram

- Population size is 1,70,450 as per 20th Livestock Census (2019).

13. Sonadi

- Sonadi breed of sheep are found in **Udaipur, Dungarpur, Banswara districts and, to some extent, Chittorgarh district of Rajasthan and also extends to northern Gujarat**.
- This breed is also called Desi, Laapdi, Bhagli.
- The name Sonadi means one with **golden fibres**. Desi means indigenous that is native to the local area, laapdi means long and flat eared and Bhagli means the lucky one as it contributes to the livelihood of the sheep farmers.
- The Animals are fairly well built, somewhat smaller than Malpura, with long legs. Light brown face with the colour extending to the middle of the neck.
- Ears are large, flat and drooping but generally have a cartilaginous appendage.
- Weight of ram: Approximately 40.4 kg and Weight of ewe: Approximately 28.5 kg.
- Tail is long and thin. The fleece is white, extremely coarse and hairy. Belly and legs are devoid of wool.
- The overall least square means of fleece weight for first six monthly and adult six monthly were 528 and 417gm, respectively. The least square means of birth, 3, 6, 9 and 12-month's weight of lambs were 2.52, 10.78, 14.98, 17.32 and 21.76 kg, respectively under farm conditions.

Fig. 20: Sonadi sheep breed: ewe and ram

- Population size is 39,373 as per 20th Livestock Census (2019).

III. Southern region

1. Nellore

- Hailing from the **Nellore, Ongole, and Prakasam districts of Andhra Pradesh**, the Nellore sheep breed is **renowned for its resilience and adaptability to tropical conditions**. These sheep exhibit strong resistance to diseases and parasites, making them well-suited to India's challenging climatic conditions.
- **Three varieties are distinguished, primarily on the basis of colour: "Palla", completely white or white with light brown spots on head, neck, back and legs; "Jodipi", white with black spots, particularly around the lips, eyes and lower jaw, but also on belly and legs and "Dora", completely brown.**
- Known for their meat quality, Nellore sheep are a favourite among meat producers and farmers seeking hardy and productive animals.

Some key features are

- **These are tall sheep with little hair** except at withers, brisket, and breech.
- They have drooping and long hair.
- Most of the Nellore sheep carry wattles.
- The male of Nellore sheep has an average body of 36 kg, and the female sheep 28 kg.

- Nellore is **the tallest sheep breed in India.**
- **They resemble goats in appearance (Sheep like goats). It has a long face and long ears with the body densely covered with short hair.**
- **Population size is 1,40,43,835 as per 20th Livestock Census (2019). This breed ranks first among all goat breeds of India having 20 % population.**

Fig. 21: Nellore sheep breed: ewe and ram

2. Deccani

- The Deccani sheep breed derives its name from the **Deccan plateau, which corresponds to its original spread across the semi-arid regions of Telangana, parts of Andhra Pradesh, Northern Karnataka, Maharashtra, and parts of Northern Tamil Nadu.**
- This breed has a **thin neck, narrow chest, prominent spinal processes. It has Roman nose and dropping ears.**
- Different strains (or within breed types) are observed in the breed tract.
- **Colour: Predominantly black or black with white markings**; some are white and brown/fawn. The distribution of colour is approximately 54.92% completely black, 21% black with white or brown spots on the head, and 24% black or white with brown spots.
- Four types of this breed have been noticed. Local people named them as **Lonand, Sangamneri, Solapuri (Sangola) and Kolhapuri.**
- **Horns: Rams have horns, while ewes are polled.**
- **Wool Colour: Different shades of black.**

- **Ear Size: Categorized as "Mulli/gundri chevulu" (very small), "Bingi/butti chevulu" (small), "Padigala/radi/solanti chevulu" (medium), and "Lambada chevulu" (large).**
- **Body Patches:** These include "**Chukka**" (small white patch on the forehead), "**Bolli**" (large white patch on the forehead), "**Batta**" (very large patch on the body), and "**Gimmala**" (patch on the flank).
- **The Deccani sheep is primarily reared for lamb production and is known locally in Telangana as the "Nalla Gorre" or Black Sheep**. Despite its resilience to climate and adaptation to semi-arid tracts, this breed is highly threatened and rapidly disappearing due to crossbreeding with non-wool hairy mutton sheep breeds.
- Means of body weights at birth, -3, -6, -9 and -12 months of age were 3.13, 14.30, 18.20, 20.10 and 22.57 kg, respectively for Deccani sheep maintained under network project.
- Sheep were sheared two times a year and the overall means for lambs first six-monthly clip and second six-monthly clip and adult annual clips were 467, 499 and 488 g, respectively.
- Deccani lambs were put on feedlot trails (90 days) after weaning with an initial average weaning weight of 13.23 kg. Final weight at six-months of age was 22.65 kg.
- Average gain of 9.13 kg was obtained with a with a feed efficiency ratio of 1: 5.91. Results shows that the Deccani has the great potential for mutton production under intensive system of management.
- Population size is 23,83,932 as per 20th Livestock Census (2019). **This breed ranks fourth among all goat breeds of India having 3.4 % population.**

Fig. 22: Deccani sheep breed: ewe and ram

3. Hassan

- The **Hassan sheep breed, as the name suggests, hails from the Hassan district of Karnataka, India**. They are distributed in the southern parts of Karnataka.
- Size: They are small to medium-sized animals.
- Colour: They have a white body with light brown or black spots.
- Ears: Their ears are medium-long and drooping.
- **Horns: Approximately 39% of the males in this breed have horns, while females are usually polled**.
- Purpose: The Hassan sheep primarily serve as a meat production breed in and around the Hassan district of Karnataka.
- They are among the five native sheep breeds **(Bannur, Deccani, Bellary, Kenguri, and Hassan)** of Karnataka recognized by the Indian Council of Agricultural Research (ICAR).
- This breed is slightly different from the Mandya breed with body colour and shape being similar to Bannur.
- The **head and neck are black or brown**. The legs are long. The quality of wool is coarse rough and many times is not useful for preparing blankets.
- The meat is less tasty compared to the Mandya breed and the dressing percentage is 45-48% with an average body weight of 30-35 kg in males and 25 - 30 kg in females.
- Population size is 7,37,401 as per 20th Livestock Census (2019).

Fig. 23: Hassan sheep breed: ewe and ram

4. Bellary

- These sheep are prevalent in **Bellari and Davangere districts** and adjoining areas of Haveri and Chitradurga districts of Karanataka.
- They are **similar to Hassan and Deccani breed** in body weight and shape.
- Their colours are black, ash like and black with white colour patches.
- The legs are lengthy which helps animals walk several miles in search of food and water without much strain.
- Rams have horns and attain 35-45kg weight in 2-2 ½ years of age, whereas ewes are polled and gain 30-35kg in 2 years.
- These animals produce low quality hairy wool fit for rough blankets. Meat is of less taste.
- Population size 42,75,218 as per 20th Livestock Census (2019). **This breed ranks second among all goat breeds of India having 6.1 % population.**

Fig. 24: Bellary sheep breed: ewe and ram

5. Kenguri

- This breed is also known as Tenguri. They are distributed throughout the hilly tracts of **Raichur district** (particularly Lingasagar, Sethanaur and Gangarati taluks) **of Karnataka**.
- They are medium-sized animals. The body coat is dark brown or of coconut colour.
- A small number of Kenguri sheep have black belly and are known as Jodka. A few sheep, known as Masaka, are a mixture of brown and black colour.

- Ears are medium long and drooping. Males are horned; females are generally polled.
- **Kenguri sheep are maintained for mutton production**.
- Lactation length varies from 3 to 5 months, and the peak milk production is 300 to 500 ml/day.
- Population size is 12,86,284 as per 20th Livestock Census (2019).

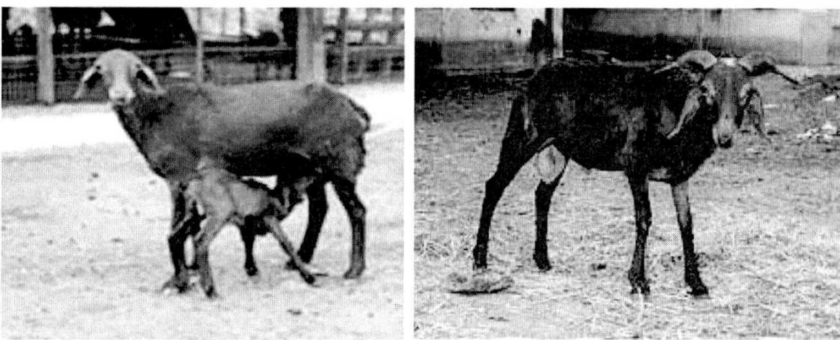

Fig. 25: Kenguri sheep breed: ewe and ram

6. Mandya

- This breed is also called Bandur or Bannur. It is renamed as Mandya breed.
- **They are bred largely in Mandya, Malavalli, and Bannur areas of Mandya district.**
- They are also found in **Mysore and Bengaluru districts.**
- The sheep have a creamy or white body colour with specific brown patches on the head and neck region.
- They **have round and broad chests with small legs**. They have bulged Roman noses and do not possess horns.
- They have a compact body resulting in higher body weight and meat yield. **Mandya breed produces meat of high quality with intramuscular fat (marbling), which makes meat soft and tasty.**
- During cooking it gives a pleasant flavour, which is a unique feature of this breed. The meat is of good quality and tasty and the dressing percentage ranged from 40-45% of the live body weight.

- This breed is also reared in other parts of India as well as abroad for developing good meat breeds of sheep.
- The sheep attains 16-18 kg live weight in 6 months and will become ready for breeding within 13-17 months of their age.
- The sheep generally produces three lambs crops in a span of two years. The adult sheep attains 35 kg body weight.
- **Mandya breed produces coarse quality wool, which is used by local peoples to produce rough blankets and carpets.**
- Population size is 2,50,038 as per 20th Livestock Census (2019).

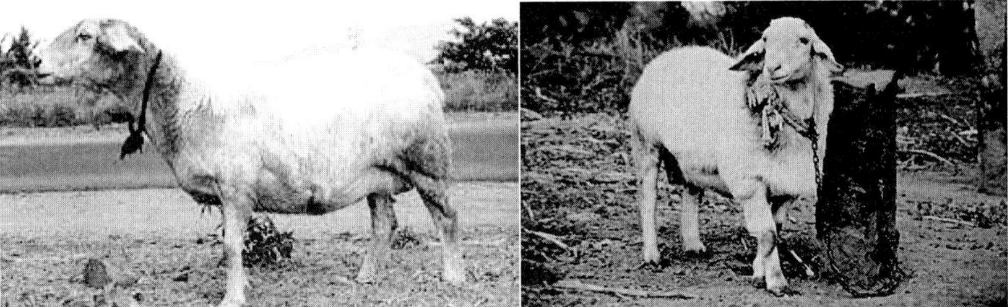

Fig. 26. Mandya sheep breed: ewe and ram

7. Coimbatore

- This breed distributed in **Coimbatore district** and adjoining **Dindigul district of Tamil Nadu.**
- Animal are medium-sized, **white with black or brown spots**.
- Ears are medium-sized and directed outward and backward.
- Tail is small and thin **Fleece is white, coarse, hairy and open**. The animals are maintained on grazing without any supplementation.
- Sheep are penned in harvested fields. **Rams are both horned and polled while ewes are hornless.**
- Average body weights at birth, 3, 6, and 12 months of age were 2.16, 7.50, 10.83 and 14.77 kg, respectively.
- Average 6-monthly greasy fleece weight was 365g with average fibre diameter of 41 micron and medullation percent of 58.

Fig. 27: Coimbatore sheep breed: ewe and ram

8. Katchaikatty-Black

- The name of this breed derives from the **Katchaikatty village** in **Vadipatti Taluk of Madurai district**, from where it originated.
- Synonyms: Katchaikatty Karuppu Aadu, Muttaadu
- The **Pallar and Konar communities** in Madurai district of Tamil Nadu have been rearing the **Katchaikatty Black sheep for mutton for decades.**
- The animals are medium in size with a black coat and the rams are especially maintained for fighting events.
- Despite its socio-economic relevance, Katchaikatty Black sheep has been listed as an **endangered breed in the National Breed Watch list.**
- The dwindling numbers of this population warrant an immediate action for its conservation.
- Population size is very less, only 1,900 as per 20th Livestock Census (2019).

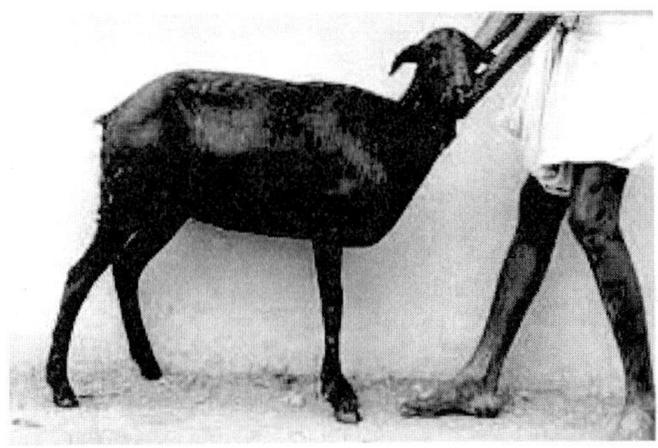

Fig. 28: Katchaikatty-Black sheep breed: ewe and ram

9. Kilakarsal

- This breed distributed in **Ramnathpuram, Madurai, Thanjavur and Ramnad districts of Tamil Nadu. They are medium-sized animals.**
- Synonyms: Keethakkaraisal, Karuvai, Keezha.
- Coat is dark tan, with black spots on head (particularly the eyelids and lower jaw), belly and legs.
- Ears are medium sized. Tail is small and thin. Males have thick twisted horns. Most animals have wattles.
- Average body weights at birth, -3, -6, and -12 months of age were 2.29, 8.53, 14.15 and 27.26 kg, respectively.
- Population size is 46,229 as per 20th Livestock Census (2019).

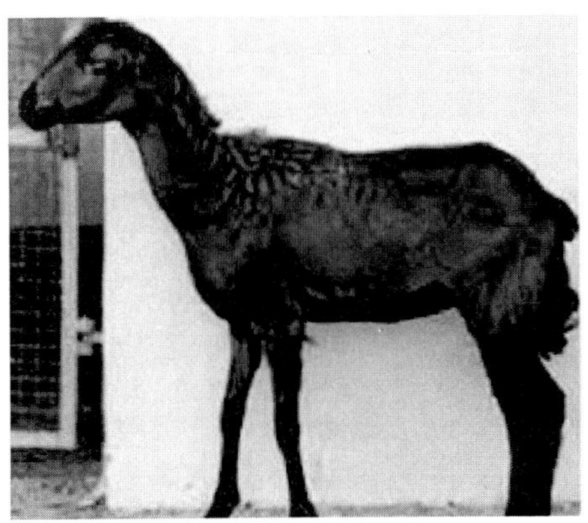

Fig. 29: Kilakarsal sheep breed: ewe and ram

10. Madras-Red

- **The Madras Red Sheep** is an indigenous breed of sheep native to the northeastern parts of Tamil Nadu (Chennai, Kancheepuram, Tiruvellore, Villupuram and adjoining areas of Vellore, Cuddalore and Thiruvannamalai districts).
- Synonyms: Chennai Sigappu.
- **Body colour is predominant brown**, the intensity varying from light tan to dark brown; some animals have white markings on the forehead, inside the thigh and on the lower abdomen.
- Ears are medium, long and drooping. The body is covered with short hairs
- Tail is short and thin

Fig. 30: Madras-Red sheep breed: ewe and ram

- Males are horned and ewes are polled. They are hardy and drought tolerant.
- Average body weights at birth, 3, 6, and 12 months of age were 2.71, 9.97, 14.84 and 21.03 kg, respectively. Dressing percentage was around 49.
- Population size is 2,97,506 as per 20th Livestock Census (2019).

1. Macheri

- **This breed distributed in Mecheri, Kolathoor, Nangavalli, Omalur and Tarmangalam Panchayat Union areas of Salem district and Bhavani taluk of Coimbatore district of Tamil Nadu.**
- Synonyms: Mailylambadi, Thuvaramchambali, Kannivadi.

- The animals medium-sized, and light brown in colour. Body is covered with very short hairs which are not shorn.
- **The skin is of the highest quality of sheep breeds in India and is highly prized.**
- Average body weights at birth, -3, -6, and -12 months of age were 2.23, 10.07, 13.73, 18.62 and 21.12 kg for male lambs and corresponding value for female lambs were 2.28, 9.70, 13.53, 15.84 and 18.02 kg, respectively under farm conditions. The corresponding growth performances recorded in farmers' flocks were 2.21, 9.64, 13.25, 16.54 and 18.72kg, respectively.
- Population size is 12,42,742 as per 20th Livestock Census (2019).

Fig. 31: Macheri sheep breed: ewe and ram

12. Nilgiri

Fig. 32: Nilgiri sheep breed: ewe and ram

- **Nilgiri sheep is a breed of sheep found only in Nilgiris district of Tamil Nadu State in India.**

- It is bred in the **hilly parts of Nilgiri and known for its fine wool**. A national plan has been implemented to save the Nilgiri sheep from extinction by the National Bureau of Animal Genetic Resources (NBAGR), Karnal, Haryana for its conservation.
- The animals are medium sized. Body colour is white; exceptionally few animals have seen with brown patches on face and body.
- Face line is convex giving a typical roman nose. Ears are broad, flat and drooping. Males have horn buds and scars; females are polled. The tail is medium and thin. The fleece is fine and dense.
- **Nilgiri** breed was evolved during the 19th century. It originated from **a crossbred base and contains an unknown level of inheritance of Coimbatore, the local hairy breed, Tasmanian Merino, Cheviot and South Down.**
- The animals are medium sized and white, exceptionally with brown patches on face and body.
- Face line is convex, giving a typical Roman nose. The ears are broad, flat and drooping.
- Males have horn buds and scurs; females are polled. Attempts have been made by Sheep Breeding Farm, Sandynallah to improve these sheep through crossing with Rambouillet and Russian merino, which has resulted in substantial improvement in fleece weight and quality.
- Means of body weight at birth, 3, 6, 9 and 12 months of age were 2.74, 9.64, 14.06, 17.04 and 19.44 kg, respectively. Average six-monthly wool yields were 438g with fibre diameter of 24.3μm and staple length of 4.5cm.
- Population size is 3,144 as per 20th Livestock Census (2019).

13. Ramanad-white

- **This breed is distributed in Ramnad district and adjoining areas of Tirunalveli district of Tamil Nadu.**
- **Synonyms: Kilakathai adu.**
- They are medium-sized animals, predominantly white. A few **animals have fawn or black markings over the body.**
- The ears are medium-sized and directed outward and downward. Males have twisted horns and females are polled.

- Tail is short and thin. Average body weights at birth, 3, 6, 12 months of age were 1.68, 7.31, 8.45 and 16.30 kg, respectively.

Fig. 33: Ramanad-white sheep breed: ewe and ram

- Population size is 3,96,856 as per 20th Livestock Census (2019).

14. Tiruchi black

- This breed is distributed in Perambalur, Ariyalur and Tiruchy, Kallakurichy taluk of South- Arcot district, Viraganur area of Attur taluk of Salem district, Tirupathur and Tiruvannamalai taluks of North Arcot district of Tamil Nadu.
- Synonyms: Tiruchirappalli, Karungurumbai.
- Males are horned and ewes are polled.
- Ears are short and directed downward and forward. Tail is short and thin. The fleece is extremely coarse, hairy and open.
- Average body weights at birth, 3, 6, 12 months of age were 2.13, 9.46, 10.73 and 16.80 kg, respectively.

Fig. 34: Tiruchi black sheep breed: ewe and ram

- Population size is 21,605 as per 20th Livestock Census (2019).

15. Vembur

- This breed distributed in Vembur, Melakkarandhai, Keezhakarandhai, Nagalpuram, Kavundhanapatty, Achangulam and some other villages of Pudur Panchayat Union and Vilathikulam Panchayat areas of Tirunelveli district of Tamil Nadu.
- Synonyms: Karandhal.
- The animals are tall. Colour is white, with **irregular red and fawn patches all over the body**. Ears are medium-sized and drooping.
- Tail is thin and short. Males are horned and ewes are polled. The body is covered with short hairs which are not shorn.
- Average body weights at birth, 3, 6, 12 months of age were 1.97, 8.42, 10.50 and 16.50 kg, respectively.
- Population size is 1,02,134 as per 20th Livestock Census (2019).

Fig. 35: Vembur sheep breed: ewe and ram

16. Chevaadu

- The Chevaadu sheep population is considered as the cultural identity of southern Tamil Nadu people. Mamman Kida and Kida Vettu are the two important cultural events / festivals of southern Tamil Nadu and the people prefer only Chevaadu rams for these festivals. Considering the significant role of this sheep breed in the livelihood of rural communities for meat, Tamil Nadu Veterinary and Animal Sciences University (TANUVAS) played a significant role in the elevation of Chevaadu to a level of a registered breed.
- **Synonyms: Arichevaadu**
- Predominantly distributing areas are Alangulam and Pappakudi panchayat union areas in Alangulam taluk; Tirunelveli taluk; Sankarankovil taluk; Nanguneri taluk of **Tirunelveli district of Tamil Nadu.**
- Two colour variants are **Light brown and Dark Brown or tan are present**.

Fig. 36: Chevaadu sheep breed: ewe and ram

- Horns are curved horizontally outward, backward with blunt conical apex having few thick ridges.
- Horns are light brown in colour, curved horizontally outward, backward with blunt conical apex having few thick ridges.
- The age at first breeding for male sheep was nine months, whereas it was 12 months in females. The oestrous cycle varies from 21-22 days with an oestrous duration of 35 hours. The age at first lambing was about 18 months. The lambing interval and gestation length were 258 and 148 days, respectively.

- Chevaadu sheep lambed thrice in an interval of two years. Litter size was recorded mostly single with a rare case of twinning.
- Population size is 67,873 as per 20th Livestock Census (2019).

17. Macherla

- Phenotypic characterization and evaluation of Macherla sheep population was done by undertaking surveys in **Guntur and Prakasham districts of Andhra Pradesh**.
- Macherla animals are medium to large in size, **mainly in white coat colour with large brown or black patches in the body, face and legs.**
- The animals with **brown patches are known as Macherla brown in breeding areas**. Convex nasal bridge, leafy ears, black muzzle and black/brown and white coat colour are predominant.
- **Males are horned and females are polled, however some females are noticed with horns in a few flocks**.
- The orientation of horns are backward, downward and forward. Ears are medium to large in size and leafy/semi-drooping. Tail is very small and thin.
- The adult body weight ranged from 38 to 69 kg in rams and 25 to 60 kg in ewes. The overall body length, height, chest girth, face length, face width, ear length and tail length were 70.07±0.32, 77.05±0.31, 87.75±0.41, 25.46±0.21, 10.49±0.08, 16.11±0.09 and 9.67±0.15 cm, respectively.

Fig. 37: Macherla sheep breed: ewe and ram

(IV) Eastern Region breeds

1. Tibetan

- The Tibetan breed is found in northern Sikkim and Kameng district of Arunachal Pradesh.
- Animals are medium sized, mostly white with black or brown face; brown and white spots are also found on the body.
- Both the sexes are horned. The nose is convex, giving a typical **Roman nose.**
- **The ears are** small, broad and drooping.
- **The fleece is relatively fine and dense**. The belly, legs and faces are devoid of wool. Average six-monthly greasy fleece yield ranges between 400-900 g. Wool parameters like staple length is 7.24 cm, fibre diameter is 19.3 μm with medullation percent of 13.22.
- Population size is 318 as per 20th Livestock Census (2019). Its population size is too small, it is at a critical level as population size is degraded.

Fig. 38: Tibetan sheep breed: ewe and ram

2. Shahabadi

- **Shahabadi is found in Shahabad, Patna and Gaya districts of Bihar/ Jharkhand state**. The animals are of medium size and leggy.
- The ears are of medium size and drooping. The tail is extremely long and thin. Both sexes are polled. Their legs and belly are devoid of wool.
- The fleece is mostly grey and sometimes with black spots. Fleece is extremely coarse, hairy and open. Average six-monthly greasy fleece yield is 240g. Fibre diameter is 49.83 μm with medullation percent of 87.08.
- Population size is 60,295 as per 20th Livestock Census (2019).

Fig. 39: Shahabadi sheep breed: ewe and ram

3. Chottanagpuri

- This breed is distributed in **Chottanagpur, Ranchi, Palamau, Hazaribagh, Singhbhum, Dhanbad and Santhal Parganas of Jharkhand, and Bankura district of West Bengal.**
- Synonyms: Gareri.
- Animals are small-sized, **lower in weight with light grey and brown colour**. Ears are small and parallel to the head. Tail is thin and short. Fleece is coarse, hairy and open and is generally not clipped.
- Population size is 3,15,069 as per 20th Livestock Census (2019).

Fig. 40: Chottanagpuri sheep breed: ewe and ram

4. Balangir

- This breed is spread over north western districts of Orissa, **Balangir, Sambalpur and Sundargarh.**
- They are medium sized, white or light brown or of mixed colours. A few animals are black. The ears are small and stumpy.

- Males are horned and females are polled.
- The tail is of medium length and thin. The legs and belly are devoid of wool. Fleece is extremely coarse, hairy and open.
- Population size is 80,435 as per 20th Livestock Census (2019).

Fig. 41: Balangir sheep breed: ewe and ram

5. Ganjam

- Ganjam sheep are found in **Koraput, Phulbani and parts of Puri districts of Orissa**. The animals are medium sized with coat colour ranging from brown to dark tan.
- Some animals have white spots on the face and body. Ears are of medium size and drooping. The nose line is slightly convex.
- The tail is of medium length and thin. While the males are horned, the females are polled.
- The fleece is hairy and short and not shorn. An annual lambing percentage of 83.6 and mortality percentage of 10.35 have been recorded in farmers' flocks.
- Under the network project on sheep improvement, a field-based Ganjam unit was started in 2001 at Orissa University of Agriculture and Technology, Bhubneswar for mutton production.
- Population size is 61,479 as per 20th Livestock Census (2019).

Fig. 42: Ganjam sheep breed: ewe and ram

6. Kendrapada

- Sheep is identified as another prolific sheep of India after Garole of West Bengal. It is the second sheep breed of India which carry FecB mutation.
- Kendrapada is distributed in **Bhadrak, Konark and Puri district of coastal area of Orissa**.
- Synonyms: Deshi, Kuzi.
- A survey was made to study the prolificacy of **Kendrapada sheep of Orissa around Kendrapada district and Nimapara (Konark).**
- The survey reveals that Kendrapada is a prolific and excellent medium stature meat type sheep.
- In the flocks surveyed more than 75% ewes produce multiple births and the adult body weight of sheep was about 23 kg.
- The coat colour varies from white to dark brown. Some black animals and black/ white spot-on bodies are also found during the survey.

Fig. 43: Kendrapada sheep breed: ewe and ram

- Population size is 10,974 as per 20th Livestock Census (2019).

7. Bonpala

- This breed is found in southern Sikkim.
- Synonyms: Banpala, Gharpala.
- The animals are **tall, leggy and well built. Fleece colour ranges from completely white to completely black with a number of intermediary tones**.
- Ears are small and tubular. Both sexes are homed. The tail is thin and short. The belly and legs are devoid of wool.
- Fleece is coarse, hairy and open.
- Average six-monthly greasy fleece yield is 500g. Fibre diameter is 49 μm with Medullation percent of 70.5.
- Population size is 68,831 as per 20th Livestock Census (2019).

Fig. 44: Bonpala sheep breed: ewe and ram

8. Garole

- Sheep are reputed for multiple births and are found in the **Sunderban area of West Bengal. These sheep are reported to have contributed prolificacy to the Booroola Merino sheep.**
- Synonyms: Meda, Bheda.
- Sundarban area has a sheep population of about 0.16 million.
- The breed tract is tropical humid. The climate of the area is hot and humid.

- Garole Wool is for rough carpet use. The average fibre diameter, medullation, staple length and crimp/cm were 67.82 μm, 75.17%, 5.09 cm and 2.08, respectively.
- **Litter size at first lambing is two and in subsequent lambing is 2 to 3. Prolificacy reported is singles 25-30%, twins 55-60%, triplets 15-20% and quadruplets 1-2%.**
- Population size is 73,582 as per 20th Livestock Census (2019).

Fig. 45: Garole sheep breed: ewe and ram

Exotic sheep breeds

1. Merino

- Merino sheep are famous for their **exceptionally soft and fine wool**, which was first brought to Spain in the late 19th century and has since become one of the world's most popular wool breeds.
- It originated in Spain and it is highly prized for its wool quality. The modern Merino sheep were domesticated in Australia and New Zealand.
- Commonly used in industrial settings such as clothing, bedding, and upholstery, Merino wool blankets are incredibly warm and cozy to sleep under on chilly winter nights.

Fig. 46: Merino sheep breed: ewe and ram

2. Rambuillet

- The Rambouillet sheep is a breed of domestic sheep from France. It is also known as the French Merino and the Rambouillet Merino. In the year of 1786, development of this breed started when Louis XVI purchased over 300 Spanish Merinos.
- As a large sized animal, average live mature body weight of the Rambouillet rams is between 113 and 135 kg. The average live body weight of the mature ewes vary from 68 to 90 kg.
- The **Rambouillet sheep are dual-purpose animals**. They are raised for both meat and wool production.

Fig. 47: Rambouillet sheep breed: ewe and ram

3. Dorset

- Wale's Horned Sheep and Spanish Merino sheep were crossed to establish the original line. But the Dorset has been around as a distinct breed for years. Today Dorset can be found in Wales, Dorset, Devon and Somerset. It's a breed that has become popular on commercial and hobby farms around the globe.
- Both horned and polled Dorset are an **all-white sheep of medium size having good body length and muscle conformation to produce a desirable carcass.**
- The fleece is very white, strong, close and free from dark fibre. Dorset fleeces average five to nine pounds (2.25-4.0 kg) in the ewes with a yield of between 50% and 70%.
- The sheep has a characteristic wide face and long body and comes in white. They can achieve the weight up to 275 lbs.
- There are two types of Dorset Sheep, the Polled and Horned Dorset. Besides the absence of horns in the Polled strain, they are identical.

Fig. 48: Dorset sheep breed: ewe and ram

4. Suffolk

- Suffolks are prolific, early maturing sheep with **excellent mutton carcasses**. They are energetic, and the whole carriage is alert, showing stamina and quality. **The breed is not desirable for wool production. The fleeces are short in staple and light in weight, and they have black fibres.**
- This is a breed of domestic sheep **from the United Kingdom**. It was originally developed in England as the result of **crossing Southdown rams on Norfolk Horned ewes.**
- The result of this cross was an improvement over both parent sheep breeds. It is a black faced, and open-faced breed and it is raised mainly for meat production.
- The **Suffolk was a recognized breed as early as 1810,** and the flock book was not closed until much later.
- **Average live body weight of the mature Suffolk rams is between 110 and 160 kg and average live body weight of the mature ewes vary from 80 to 120 kg.**

Fig. 49: Suffolk sheep breed: ewe and ram

5. Corriedale

- The Corriedale was developed in New Zealand and Australia during the late 1800's from **crossing Lincoln or Leicester rams with Merino females.**
- The Corriedale is of medium to large size; grown ewes weigh some 65 – 75 kg, full-grown rams 85 – 105 kg. It is polled, white-woolled and white-faced, with dark hooves and dark skin on the nostrils.
- Ewes have good maternal qualities but are not highly prolific – the twinning rate is in the range 5–25%.
- **The Corriedale produces bulky, high-yielding wool ranging from 31.5 - 24.5 micron fibre diameter.**
- The fleece from mature ewes will weigh from 10 - 17 pounds (4.5 - 7.7 kg) with a staple length of 3.5 - 6.0 inches (9.0 - 15 cm).
- The yield percent of the fleece ranges from 50-60%.

Fig. 50: Corriedale sheep breed: ewe and ram

6. Karakul

- The **Karakul** sheep are named after a village called Karakul, which lies in the valley of the **Amu Darja River in the former emirate of Bokhara, West Turkestan**.
- Between 1908 and 1929, the Karakul sheep were first introduced to the United States. In USA, this breed has been an increased interest of farmers, with a growing interest in the fibre arts in the nation.
- **Karakul sheep are a multi-purpose breed, kept for milking, meat, pelts, and wool. As a fat-tailed breed, they have a distinctive meat quality.**
- Karakuls are medium-sized sheep. The rams will weigh between 175-225 pounds and the ewes range from 100-150 pounds. They stand tall, with a long, narrow body. The top line is highest at the loin with the rump long and sloping, blending into a low set broadtail.

Fig. 51: Karakul sheep breed: ewe and ram

Crossbreeding of sheep in India

1. **Hissardale:** Hissardale was evolved at the Government Livestock Farm, Hissar, through crossbreeding of **Australian Merino** rams with

Bikaneri (Magra) ewes and stabilizing the exotic inheritance at about 75%. There is a small flock of this breed at the Government Livestock Farm, Hissar. The rams of this breed were earlier distributed in the hilly regions of Kullu, Kangra etc.

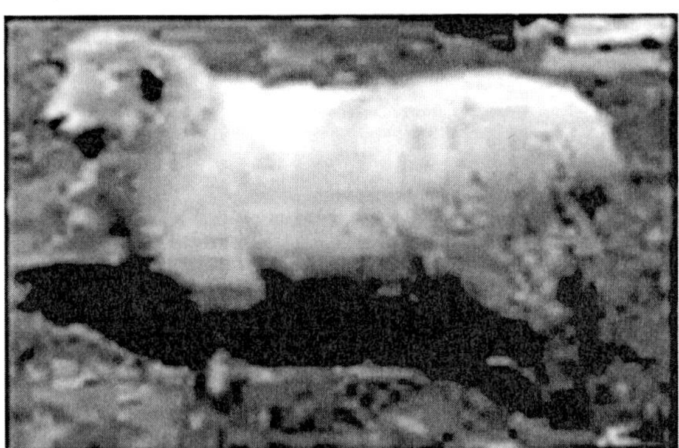

Fig. 52: Hissardale sheep crossbreed

2. **Harnali:** Harnali sheep is a three-breed cross by involving **Nali** (37.5%) and exotic (62.5%) inheritance of **Merino and Corriedale** with equal inheritance, i.e., (31.25%) developed at Lala Lajpat Rai University of Veterinary and Animal Sciences, Hisar for superior wool production.

Fig. 53: Harnali sheep crossbreed: ewe and ram

3. **Kashmir Merino:** This breed originated from crosses of different Merino types with predominantly migratory native sheep breeds such as **Gaddi, Bhakarwal and Poonchi**. The level of inheritance varies from **very low to almost 100% Merino, though a level from 50 to 75% predominates.** The animals are highly variable because of the involvement of a number of native breeds. The annual greasy-fleece weight (kg) is 2.80 having average fibre diameter of 20.4μm.

Fig. 54: Kashmir Merino sheep crossbreed: ewe

4. Avikalin

The Avikalin strain has been evolved by the crossing of **Rambouillet x Malpura.** Half-bred base through inter-breeding and selection for greasy fleece weight, producing about 1.75 kg greasy wool having 27 microns diameter, 27% medullation and staple length of 4.75 cm in half yearly clip. This breed is quite suitable as a dual-purpose sheep for carpet wool and mutton production.

Fig. 55: Avikalin crossbreed sheep

5. Avishaan

ICAR-Central Sheep & Wool Research Institute, Avikanagar released a **high performing prolific three breed cross** developed on 04.01.2016 on the occasion of Institute foundation day for testing in the farmers' field is **'Pride of CSWRI Avishaan: a prolific sheep'.**

- The prolific cross is a composite cross having **12.5% Garole, 37.5% Malpura and 50% Patanwadi blood.**
- The average growth performance of "Avishaan" in farm condition is 3.30 kg at birth, 16.80 kg at 3-months, 25.90 kg at 6-months and 34.70 kg at 12-months age.

- “Avishaan” excelled by about 30% over native Malpura sheep in terms of Ewe Productivity Efficiency (EPE) at 6-months of age.
- **In the current female population, 47% females produced 2 lambs and 2% produced 3 lambs in a lambing leading to overall 49% prolificacy.**
- With an enhanced growth, farmers can get 1.5 times income from rearing of “Avishaan” without an extra investment. The enhanced milk production will lead to sustenance of more lambs, fulfilment of domestic requirement, thus adding up to the nutritional security of the sheep farmers.
- During the occasion five elite “Avishaan” breeding rams along with data recording aids were distributed to five registered farmers of five villages around Malpura for extensive field evaluation for validation of the genotype.

Fig. 56. Avishaan crossbreed ewes with triplet and twins in above pictures.

Crossbreds of sheep developed in the country

Sl. No.	Crossbred sheep	Utility	Exotic breed	Indian breed	Exotic inheritance (%)
1	Avikalin	Carpet wool	Rambouillet	Malpura	50
2	Avivastra	Fine wool	Rambouillet	Chokla	50
3	Avimaan	Mutton	Dorset and Suffolk	Malpura and Sonadi	50
4	Hissardale	Apparel wool	Australian Merino	Bikaneri	75
5	Kashmiri Merino	Fine wool	Merino and Rambouillet	Gaddi, Bhakarwal and Poonchi	50-75
6	Chokla synthetic	Carpet wool	Merino and Rambouillet	Chokla	50
7	Nali Synthetic	Carpet wool	Merino and Rambouillet	Nali	50
8	Nilgiri Synthetic	Apparel wool	Merino	Nilgiri	62.5-75
9	Patanwadi synthetic	Carpet wool	Merino and Rambouillet	Patanwadi	50
10	Bharat Merino	Fine wool	Merino and Rambouillet	Chokla, Nali, Malpura and Jaiselmeri	75
11	Nellore Synthetic	Mutton	Dorset and Suffolk	Nellore	50
12	Mandya Synthetic	Mutton	Dorset and Suffolk	Mandya	50
13	Indian Karakul	Pelt	Karakul	Marwari and Malpura	75
14	Harnali	Superior wool production	Merino and Corriedale with equal inheritance, i.e., 31.25%	37.5% Nali	62.5 exotic

4

Goat Breeds of India

In India, goats are known as "poor man's cow" because they are an important source of income for poor and landless people. They can be reared on cheaper resources easily than other livestock.

Goats come under the genus *Capra* have evolved from five wild species, including *Capra hircus* (which includes Bezoar), *Capra ibex*, *Capra caucasica*, *Capra pyrenaica*, and *Capra falconeri*. This evolutionary pattern contributes to their phenotypic diversity and adaptability to various environments. The chromosome profile of goats (specifically the *Capra hircus* species) reveals some interesting features which are as follows:

Chromosome Number

The modal chromosome number for goats is 60.

These chromosomes can be categorized into different types based on their centromere position:

Autosomes: These are either acrocentric or telocentric.

X Chromosome: It is acrocentric.

Y Chromosome: It is submetacentric.

Identification

Quinacrine-mustard staining and fluorescence microscopy allow us to identify 28 out of 30 chromosome pairs in goats.

The X chromosome is one of the largest chromosomes, while the Y chromosome is the smallest.

Q-Band Pattern

The Q-band pattern of the autosomes in goats seems to be identical to that of cattle. However, the sequence of Q-bands on the X chromosomes differs between goats and cattle.

Important features

- About 97% of the world's goats' population are found in Asia, Africa and Latin America. Asia accounts for the largest share, with approximately 51% of the total.
- Most dairy goats are raised in the Mediterranean region, South Asia and parts of Latin America and Africa.
- Major goat milk producer countries are India, Bangladesh and Sudan.
- The average milk yield of various goat breeds varies significantly in the major milk producing countries. In Sudan, the average goat milk yield is about 64 kg/year, while in India it is more than 179 kg/year.
- Milk from goats contributes significant shares of total milk production in sub-Saharan Africa (9%) and parts of South, East and Southeast Asia (excluding China).
- The countries with the majority of dairy goats are India, Bangladesh and Mali (FAO).
- Boer, Spanish, Kiko, Pygmy, Black Bengal, Sirohi, Frisian etc. are some meat producing goat breeds and are suitable for stall fed goat farming system.

Goats produce higher milk than sheep. They are considered the "**poor man's cow**" and are the major source of milk and meat for many subsistence farmers in tropical regions. They are commonly reared in the arid and semi-arid areas and are generally kept in small flocks of two to ten animals. Goat milk is widely produced in West Africa but also in the Caribbean and Central Africa, usually for household consumption, although it is sometimes traded within the community. Compared with that of dairy cows, the lactation curve of goats is flatter, with a less prominent peak and greater persistency. In some cases, the lactation curve may have two peaks because of seasonal fluctuations in feed availability.

Although the majority of dairy goats are found in developing countries, the breeding programmes are concentrated in Europe and North America. Genetic selection of dairy goats has resulted in considerable increases in yield and longer lactation length. The specialized dairy goat breeds used in developed countries, therefore, have higher genetic potential for milk production than the breeds found in the developing countries. In the past decades, many specialized breeds have been exported to various developing countries for crossing with local breeds to improve milk production. The most widely distributed dairy goat breeds are Saanen, Anglo-Nubian, Toggenburg, Alpine and West African Dwarf (FAO,2010). In Switzerland goat is nicknamed as “Swiss baby’s foster mother”.

The Scientific Classification of Goats

Kingdom: Animalia

Phylum: Chordata

Class: Mammalia

Order: Artiodactyla

Suborder: Ruminantia

Family: Bovidae

Subfamily: *Caprinae*

Genus: *Capra*

Species: *Capra aegagrus hircus*

There are about 39 well defined Indian breeds apart from a number of local non-descript goats scattered throughout the country. The two new breeds added by NBAGR in December 2023 are Anjori Goat (Chhattisgarh) and Andamani Goat (Andaman and Nicobar Islands). All the goat breeds are classified based on their geographical locations.

I. Himalayan - Region (Hilly tract)

This region comprises the states namely Jammu and Kashmir, Himachal Pradesh, Uttarakhand and parts of Uttar Pradesh.

Himalayan breed: These goats are white haired and sturdily built. The breed is also known as Gaddi, Jamba, Kashmiri according to their localities where they are reared. They inhabitant Kangra and Kullu valleys, Chamba, Sirmur and Shimla in Himachal Pradesh and parts of Jammu hills. Castrated bucks are used for transporting merchandise in the hilly tracts.

Pashmina: These are small dainty animals with quick movements. They are raised above 3400 m elevations in the Himalayas, Ladakh and Lahaul and Spiti valleys. They produce the softest and warmest animal fibre used for high quality fabrics. The yield of pashmina varies from 75-150 g/goat.

Chegu: This breed is found in the mountainous range of Spiti, Yaksar and Kashmir. The goats of this breed produce pashmina, good meat and a small quantity of milk.

Table 1: Goat breeds of Himalayan region of India

Sl. No.	Breeds	State	Use
1	Chegu	Jammu and Kashmir	Fibre called Pashmina
2	Changthangi	Jammu and Kashmir	Fibre called Pashmina
3	Gaddi	Himachal Pradesh	Fibre
4	Bhakarwali	Himachal Pradesh	Fibre

II. Northern Region

The states that come under this region are Punjab, Haryana, Uttarakhand and parts of Uttar Pradesh. The important milch breeds of goats and meat too are distributed in this region.

Table 2: Goat breeds of northern region of India

Sl. No.	Breeds	State	Use
1	Beetal	Punjab	Meat and milk
2	Jamunapari	Uttar Pradesh	Milk
3	Rohilkhandi	Uttar Pradesh	Meat and milk
4	Barbari	Uttar Pradesh and Rajasthan	Meat and milk
5	Pantja	Uttarakhand and Uttar Pradesh	Meat and milk

III. Central and Western regions

This region includes Rajasthan, Madhya Pradesh, Gujarat and northern parts of Maharashtra. These breeds come in different colour combinations. They are derived from the Jamunapari breed. Commonly found in Rajasthan, Marwari, Mehsana and Zalwadi. They yield between 0.75 -1.0 kg of milk per day.

Table 3: Goat breeds of central and western region of India

Sl. No.	Breeds	State	Use
1	Jakhrana	Rajasthan	Milk and meat
2	Marwari	Rajasthan	Milk and meat
3	Sojat	Rajasthan	Milk and meat
4	Karauli	Rajasthan	Milk and meat
5	Gujari	Rajasthan	Milk and meat
6	Sirohi	Rajasthan and Gujarat	Milk and meat
7	Gohilwadi	Gujarat	Milk and Meat
8	Kutchi	Gujarat	Milk and Meat
9	Mehsana	Gujarat	Milk and Meat
10	Surti	Gujarat	Milk
11	Zalawadi	Gujarat	Milk and Meat
12	Kahmi	Gujarat	Milk and Meat
13	Osmanabadi	Maharashtra	Milk and meat
14	Sangamneri	Maharashtra	Meat
15	Konkan Kanyal	Maharashtra	Milk and Meat
16	Berari	Maharashtra	Milk and Meat
17	Anjori	Chhattisgarh	Meat

IV. Southern region

The states under this region include parts of Karnataka, Andhra Pradesh, Tamil Nadu and Kerala.

Table 4: Goat breeds of southern region of India

Sl. No.	Breeds	State	Use
1	Bidri	Karnataka	Meat
2	Nandidurga	Karnataka	Meat
3	Attapady Black	Kerala	Meat
4	Malabari	Kerala	Milk and meat
5	Kodi Adu	Tamil Nadu	Meat
6	Salem Black	Tamil Nadu	Meat
7	Kanni Adu	Tamil Nadu	Meat

V. Eastern Region

This region consists of West Bengal, Assam, Meghalaya, Tripura, Nagaland, Orissa, part of Bihar. Andaman and Nicobar.

Table 5: Goat breeds of eastern region of India

Sl. No.	Breeds	State	Use
1	Teressa	Andaman and Nicobar	Meat
2	Assam Hill	Assam and Meghalaya	Meat
3	Sumi-Ne	Nagaland	Meat
4	Ganjam	Orissa	Meat
5	Black Bengal	West Bengal	Meat
6	Andamani	Middle and North Andaman Island	Meat

Exotic Breeds

The important exotic dairy breeds of goats are Toggenburg, Saanen, French Alpine and Nubian. They all are noted for their **higher milk yield and most of these breeds were imported to India to improve milk yield of our local breeds** and to upgrade our nondescript goats.

1. **Toggenburg:** It originated in **the Toggenburg valley in north Switzerland**. Skin is very soft and pliable. Usually both males and females are hornless. The adult does weighs 65 kg or more and the bucks more than 80 kg. **Average milk production is 5.5 kg per day. The butter fat content of milk 3-4%.** The males usually have longer hairs than females.

Fig. 1: Toggenburg goat breed: doe and buck

2. **Saanen:** Saanen goats, often referred to as the "**Holsteins of the dairy goat world**," or **queen of dairy goat** originate from the Saanen Valley in Switzerland. This breed is very famous for its consistency and high production. Coat colour is white or light cream. The face may be slightly dished and the ears point upward and forward. Both sexes are normally polled but sometimes horns do appear. Does weigh 65 kg and the bucks 95 kg. Average milk yield is 7.5-11.0 litre per day during a lactation period of 8 -10 months. These strikingly white goats are known for their high milk production, making them a popular choice for dairy farmers worldwide. Average milk yield is 838 kg in a lactation of 264 days. Milk fat is 3 - 5%.

Fig. 2: Saanen goat breed: doe and buck

3. **Alpine:** This breed was originated in Alp's mountains. It was derived from French, Swiss and Rock Alpine breeds. No distinct colour has been established and they have horns. They are excellent milk producer. The average milk yield is 2 - 3 kg with butter fat of 3 - 4%.

Fig. 3: Alpine goat breed: doe and buck

4. **Nubian:** This breed originated from Nubia of North Eastern Africa and also found in Ethiopia and Egypt. These are long legged and hardy animal. This breed along with Jamunapari of India together with native breeds of U.K. formed the crossbred Anglo Nubian breed of goat.

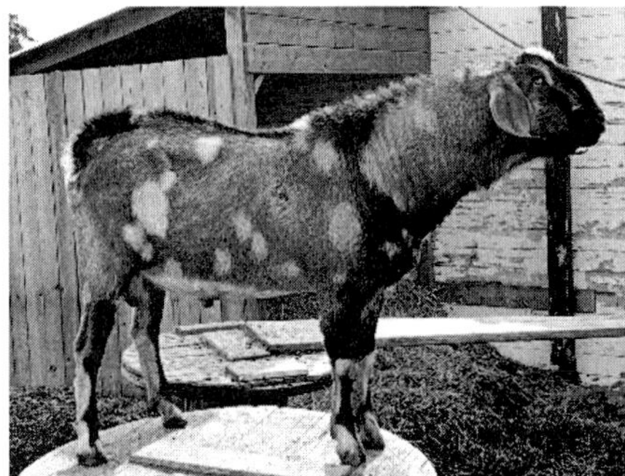

Fig. 4: Nubian goat breed: doe and buck

5. **Anglo Nubian:** This is a large sized breed of goat with a fine skin and glossy coat, pendulous ears and Roman nose. Anglo Nubian is known as the **Jersey cow of the goat world**. Udder is large and pendulous with bigger teats. There is no fixed colour. Bucks weigh 65-80 kg and does from 50-60 kg. Average milk yield in 3-4 kg/day. Peak yield may even go up to 6.5 kg or more.

Fig. 5. Anglo Nubian goat breed: doe and buck

6. **Angora:** This breed originated in **Turkey or Asia minor**. They produce a superior quality fibre called **mohair**. The soft silky hairs cover the white body. If not shorn during spring the fleece drops off naturally as summer approaches. Average fleece yield is 1.2 kg. Good animals yield even up to 6 kg. The Angora is small in size with short legs. Horns are grey, spirally twisted and inclined backward and outward. Tail is short and erect.

Fig. 6: Angora goat breed: doe and buck

7. **Boer:** This goat breed has an interesting history and origin, because it originated from the indigenous South African goats kept by the Namaqua, San, and Fooku tribes (with crossing of Indian and European bloodlines). This is a very common and popular meat goat breed in the United States and many other countries. These goats are of a much bigger size than many other popular meat goat breeds. They also produce milk, but are mostly suitable for meat production. The breed was first introduced to the United States in the early 1990s and gained popularity quickly among the farmers and ranchers mainly due to its

superior meat quality and ability to thrive in diverse environments. Boer goats commonly have white bodies and distinctive brown heads but a few can be completely brown or white or paint, which means large spots of a different colour are on their bodies. The does are reported to have superior mothering skills. Usually, a mature buck weighs about 110 to 135 kg and mature does weigh about 90 to 100 kg. Boer does usually give birth to single or twins per kidding but twins are more common. **The Boer goat is indeed considered one of the best breeds for meat production, and it's highly regarded for stall-fed farming.**

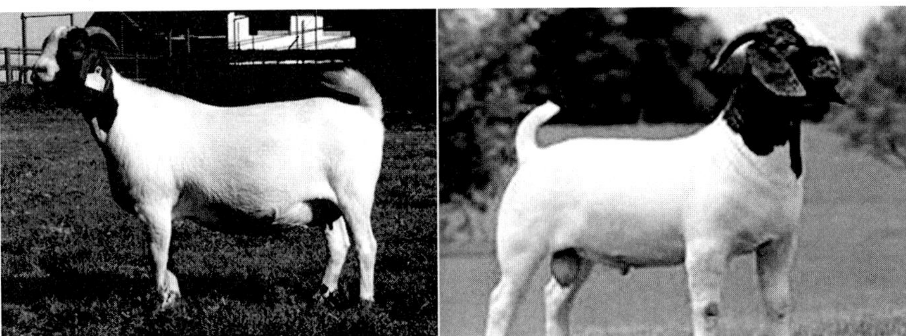

Fig. 7. Boer goat breed: doe and buck

Description of Indian breeds

I. Himalayan breeds

1. Chegu

- The Chegu is a pashmina bearing goat distributed in mountainous, cold arid region of Lahaul - Spiti and Kinnaur in Himachal Pradesh and high altitude (above 12000 feet MSL) areas of Uttarkashi, Chamoli and Pithoragarh districts of Uttarakhand bordering Tibet.
- **Synonyms: Chayangra, Pashmina.**
- **The goats of this breed yield of pashmina, good meat and a small quantity of milk.**
- Breeding tract: The breeding tract is confined to cold desert region of Spiti division (Lahaul & Spiti districts); Hangrang valley of Kinnaur district; Todd and Miar valley areas of Lahaul region; and high-altitude areas of Sural and Udeen in Pangey valley (Chamba district).
- Coat colour is predominantly **white but greyish red and mixed colours** are also seen.

- Average live bucks weigh 39 kg and does 26 kg.
- Average birth weight 2.0 kg.
- Kidding is once a year and mostly single.
- Average lactation milk yield and lactation length are 69 kg and 187 days, respectively.
- Long hairs below which under-coat delicate fibre (**cashmere or pashm**). Used for draught (pack) to carry salt and small loads.
- Population size is 2,356 as per 20th Livestock Census (2019).

Fig. 8: Chegu goat breed: doe and buck

2. Changthangi

- The Changthangi or Changpa is a breed of cashmere goat native to the high plateaus of Ladakh in northern India. It is closely associated with the nomadic Changpa people of the Changthang plateau. It is also known as the Ladakh Pashmina or Kashmiri.
- Synonyms: Kashmiri, Pashmina.
- Predominantly white and rest brown, grey and black. Undercoat white/grey; yields warm delicate fibre, pashmina (**cashmere, pashm**).
- Average age at first kidding 20 months.
- Body and legs are small in size, strong body and powerful legs.
- Average live weight of bucks 20 kg and does 20 kg; average birth weight 2.1 kg.
- Kidding is once a year, normally single.
- Pashmina yield per shearing varies depending on the method used and the type of goat. Hand-combing involves gently removing the fine

undercoat fibers, while shearing involves clipping the fleece to collect the undercoat. The average pashmina production per goat in the traditional belt is around 130-350 g per animal.

- Population size is 2,05,940 as per 20th Livestock Census (2019).

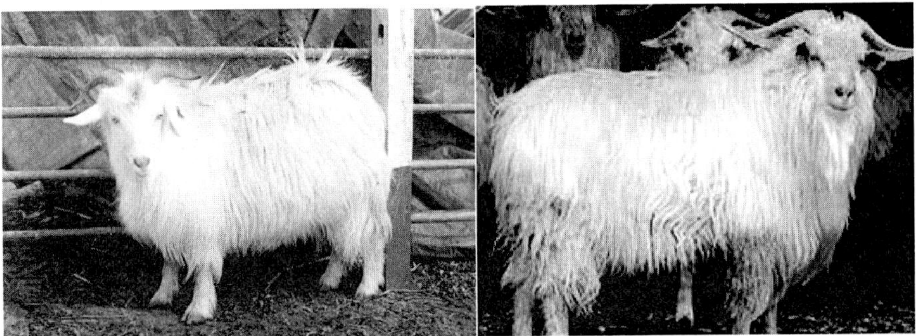

Fig. 9: Changthangi goat breed: doe and buck

3. Gaddi

- The Gaddi breed, also known as the White Himalayan, is distributed throughout **Chamba, Kangra, Kullu, Bilaspur, Simla, Kinnaur and Lahaul and Spiti in Himachal Pradesh and Dehradun, Nainital, Tehri Garhwal and the Chamoli districts of Uttarakhand**.
- Synonyms: Chamba, Gadderan, Gadhairun, Kangra valley, White Himalaya.
- They are well built and sturdy animals and have drooping and pointed ears with long pointed horns bending upwards and backwards.
- Kidding takes place once a year with single births. The hairs are about 17-19 cm long and lustrous; one shearing yield 500 g to 1.0 kg of wool.
- The meat is coarse and devoid of fat, and milk yield averages about 1.5 kg/day. According to the 18th Livestock Census, 2007, the numbers of Gaddi goats in the country are 472,405.
- Population size is 7,38,425 as per 20th Livestock Census (2019).

Fig. 10: Gaddi goat breed: doe and buck

4. Bhakarwali

- These are white coloured goats distributed in the Jammu division of Jammu and Kashmir.
- Synonyms: Kagani.
- Breeding tract: Jammu region covers a wide area ranging from a thin strip 274 m to 304 m (900-1000 ft) in height of the plain region in the south to the Shivalik hills and mid Himalayan mountains northwards up to Pir Panjal range. The region represented an intricate mosaic of mountain ranges and hills characterized with river terraces, valleys and gorges. It had special importance because of its ecology and environmental hostilities. The rich forests of the state played an important role in maintaining the ecological imbalances.
- Face or hind quarters are black in some animals. Pure black goats are also found.
- Whole body is covered with long hair. These are large sized goats having a convex head.
- Ears are cut and pendulous. Horns are screw type and are carried upwards and backwards.
- These are reared for meat and milk. Udder is pendulous.
- Adult body weight varies from 35 to 60 kg in males and 30 to 50 kg in females.
- Average daily milk yield is about 900g.

Fig. 11: Bhakarwali goat breed: doe and buck

II. Northern region breeds

1. Beetal

- This breed is found in **Punjab and Haryana** states but it is mostly found in Gurdaspur, Amritsar and Ferozpur districts of Punjab.
- Synonyms: Amritsari.
- Beetal goats are raised both for **meat and dairy purposes**. They have long legs, long and pendulous ears, short and thin tails and have backward curved horns.
- The adult buck has 50-60 kg weight and adult doe has 35-40 kg weight. The body length of male Beetal is approximately 86 cm and that of a female Beetal is approximately 71cm.
- Black coat is common. Brown with white spots of different sizes is also available.
- The average milk yield per day is 2.0-2.25 kg and yield per lactation period varies from 150-190 kg.
- Population size is 12,34,760 as per 20th Livestock Census (2019).

Fig. 12: Beetal goat breed: doe and buck

2. Jamunapari

- Jamunapari goat is also known by the name of **'Ram Sagol'** and 'Jamanapari'. Because of its majestic appearance, it is commonly known as 'Pari' in the area of its origin.
- Jamunapari goat breed is gaining popularity throughout the world due to their fast growth rate, higher milk yield, and excellent reproductive efficiency. The other things that are making them popular is their suitability to almost any kind of climate and less disease incidence than other goat breeds.
- Jamunapari is considered the **best goat breed in India for goat farming**. It is a dual-purpose breed, excelling in both milk and meat production. On average, Jamunapari goats can produce 1.5 liters of milk per day during their lactation period.
- This breed has been also used to create other breeds. **American Nubian** goats originated from Jamunapari goats.
- Jamunapari goat is a very **beautiful dairy goat** breed which originated from India. **This breed was first introduced near a river of Uttar Pradesh named Jamuna.** Since this breed is mostly known as Jamunapari goat.
- Synonyms: Etawah.
- The body of this goat is comparatively long in size than other goat breeds. They are of **white, black, yellow, brown or various mixed colours.** They have long sized legs with long hair in the thigh and back legs.
- They generally love to graze in the open field. Tails of these goats are short and typically curved upward. They have very long, thin ears which are curved downwards.
- Most of the goats have a pair of small sized horns which are curved backwards. They produce one kid per year. Does generally produce about 1.5-2.0 liters milk daily.
- They are large sized goats, tall and leggy, with a convex face line and large folded pendulous ears. They are generally found in white colours.
- Average birth weight up to 4.0 kg.
- Average age at first kidding is 20-25 months.

- They have large udders and big teats and average yield is 280 kg in 274 days lactation period
- They have the ability to yield 2.0 to 2.5 kg of milk per day.
- The fat content of the milk ranges between 3.0 to 3.5%.
- An adult female weighed between 45 kg to 60 kg, whereas an adult male ranged between 65 kg to 80 kg.
- High milk producing breeds are not found in Tamil Nadu climatic conditions.
- They thrive best under range conditions with plenty of shrubs for browsing.
- Population size is 25,55,965 as per 20th Livestock Census (2019). It is 1.7% share among total goat population in India.

Fig. 13: Jamunapari goat breed: doe and buck

3. Rohilkhandi

- Native to **Rohilkhand region of Uttar Pradesh**, and is reared for **meat and milk**.
- Synonyms: Desi, Chheri.
- Colour: Coat colour is predominantly black with star or patch on neck and face in some animals.
- Horns shape and size: Majority of animals are horned which are curved, and directed laterally and outwardly.
- Characteristics: Beard and wattles are absent in both sexes. Forehead is slightly convex. Tufts of hair (black or brown) are present in the thigh region. Tail is bunchy.

Fig. 14: Rohilkhandi goat breed: doe and buck

4. Barbari

- **Popular in urban areas of Delhi, Uttar Pradesh, Gurgaon, Karnal, Panipat and Rohtak in Haryana state.**
- Synonyms: Sai bari, Titri bari, Wadi bari, Bari.
- The Barbari goats get their name from **Berbera**, which is a coastal city located on the Indian Ocean Somalia.
- **This is a dwarf breed highly suitable for stall feeding** and generally found in cities so it is also known as "city breed".
- These types of breeds are found mostly in the states of Delhi, Uttar Pradesh and Haryana.
- Barbari breeds are grown mainly for **milk and meat purposes**.
- **The colour of this breed is a mixture of white and purple.**
- An adult female goat weighs between 25 kg to 35 kg, whereas an adult male goat ranges between 35 kg to 45 kg.
- They have the ability to give 1.0 kg to 1.5 kg of milk per day.
- This breed has better reproductive capabilities and gave birth to 2 - 3 kids in parturition. They are usually stall-fed and reported to yield 0.90-1.25 kg of milk per day (with fat content 5%) in a lactation period of 108 days. They are prolific breeders and produce kids twice in 12-15 months. Population size is 47, 59,305 as per 20th Livestock Census (2019). This breed has a percentage share of 3.2% among Indian breeds and third highest in population.

Fig. 15: Barbari goat breed: doe and buck

5. Pantja

- Pantja goat is reared for **meat and milk** in **Udham Singh and Nainital districts of Uttarakhand and adjacent Tarai area of Uttar Pradesh**.
- Synonyms: Pantja.
- Colour: Mix of light brown and white in females and males are mix of white.
- Black Horn Shape and Size: Straight oriented backwards. Small in size.
- These goats are well adapted to humid condition of Tarai region. Twining is common in Pantja goat.
- Characteristic: Long hairs on thigh region, straight horns. Goats are well adapted to humid condition of Tarai region.
- Population size is 28,728 as per 20th Livestock Census (2019).

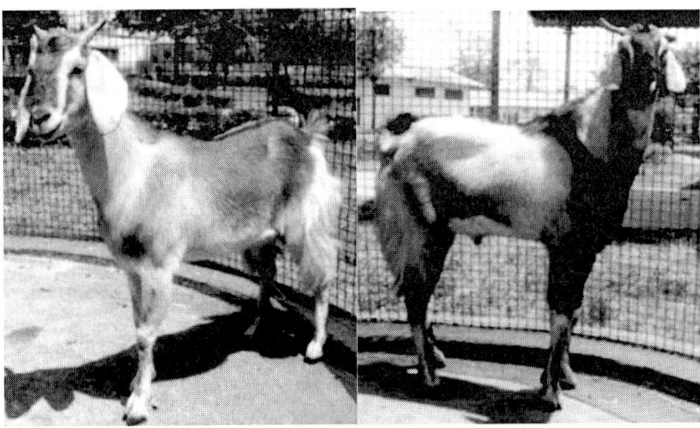

Fig. 16: Pantja goat breed: doe and buck

III. Central and Western Region

1. Jakhrana

- Home track: Jakhrana is an important dairy breed of the semi-arid tract of the Rajasthan State of India. The habitat of this breed is a small hamlet in the **Aravali hill ranges**.
- Synonyms: Kali kotri.
- The breed is spread over a limited area and the population size is small. The breed derives its name from the **Jakhrana and few surrounding villages near Behror, of Alwar district of Rajasthan where it is found in its purest form.**
- Colour: Predominantly black with white spots on ears and muzzle.
- Horn Shape and Size: Broad and flat, going backwards
- Characteristic: Straight face line. Forehead is narrow and slightly bulging. Udder size is large with conical teats.
- Animals are large and predominantly black with white spots on ears and muzzle.
- The breed is very similar to Beetal, the main contrast is that Jhakrana is comparatively longer.
- These goats are used mainly for milk production. Average daily milk yield varies from 2.0-3.0 kg for a lactation length of about 180-200 days.
- These are prolific. Kidding is mostly single but in 40% cases twins are born. The goats are also reared for meat production, and their skins are popular with the tanning industry.
- Population size is 6,55,582 as per 20th Livestock Census (2019).

Fig. 17: Jakhrana goat breed: doe and buck

2. Marwari

- The native tract of the Marwari goat breed is western Rajasthan the districts of **Barmer, Jaisalmer, Bikaner, Jodhpur, Jalore, Pali and Nagaur.**
- Synonyms: Barmeri, Black Desert goat of Rajasthan.
- The breed derived its name from Marwar region of Rajasthan, which is natural habitat of the breed
- This is a dual-purpose breed, reared for both mutton and milk, and is well adapted to the harsh environment of the Thar desert.
- The Marwari goat is a medium sized animal, **predominantly black in colour**. The hair covering is lustrous and the tail is small and thin.
- The udders are fairly well developed but small and round with small teats placed laterally.
- The milk yield varies from 0.5 to 1.0 kg when reared on grazing and from 2.0 to 3.0 kg under stall fed conditions.
- Kidding is primarily single births, with a twinning of around 10%, which increases when the goats are kept under stall fed conditions and provided with the supplementary feed.
- The thick hairs protect the animal from the extremes of temperature found in this region.
- The hairs are used to weave traditional harnesses for camels, and also carpets and bags, the latter used by potters.
- The males have a thick beard. The ears are small and flat, carried on a small head. Both sexes have short pointed horns, directed upward and backward.
- Population size is 50,41,776 as per 20th Livestock Census (2019), which is the second highest population after Black Bengal with a share of 3.4% among Indian goat population.

Fig. 18: Marwari goat breed: doe and buck

3. Sojat

- It is a large, dual-purpose goat breed, distributed over the **Sojat, Pipar, Pali, Jodhpur, Nagaur and Jaisalmer districts of Rajasthan.**
- The coat colour of these animals is **white with brown spots on head, neck, ear and legs**; however, **pure white animals are also found in the field.**
- Wattles are present in the majority of females while completely absent in males.
- Horns are absent in most of the goats of both sexes.
- Average adult weight is about 60.0 kg in males and 53.0 kg in females.
- Average daily milk yield, lactation milk yield and lactation length are 1060.12±12.59 g, 266.64±0.63 kg and 232.92±1.17days, respectively.

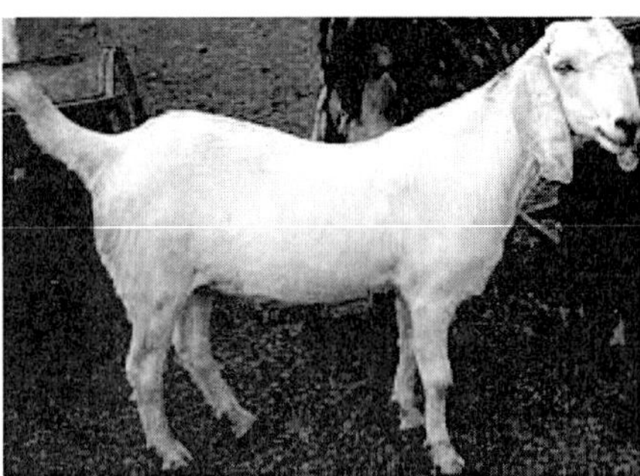

Fig. 19: Sojat goat breed: doe and buck

4. Karauli

- Karauli goats are medium to large in size and dual-purpose breed, which is distributed in **Sawai Madhopur, Kota, Bundi, and Baran districts of Rajasthan.**
- The coat colour pattern is **black with brown stripes** on face, ears, abdomen, legs and near pin bones.
- Ears in **Karauli goats are long, pendulous with folded and brown lines on the border of ears**. The animals have roman noses.
- The horns are medium sized corkscrew in shape which are pointed upwards are the most typical feature of Karauli goats.
- **Karauli bucks have prominent hanging dewlap**.
- Average adult weight is about 52.0 kg in males and 45.0 kg in females. Average daily milk yield, lactation milk yield and lactation length are 1530.43±19.61 g, 270.04±2.24 kg and 251.70±6.53 days, respectively.

Fig. 20: Karauli goat breed: doe and buck

5. Gujari

- **Gujari goat is a large, dual-purpose breed, distributed mainly in Jaipur and Sikar districts of Rajasthan.**
- The animals are **brown and white mixed coat colour, while white coloured face, leg and abdomen are typical features of the breed.**
- Ears are long, pendulous and folded, and horns are small, backward and twisted.
- Males have a beard, while it is completely absent in adult females.
- Dewlap is present in the majority of animals.

- Average adult weight is about 69.0 kg in males and 58.0 kg in females.
- Average daily milk yield, lactation milk yield and lactation length are 1616.47±11.45 g, 347.54±2.24 kg and 250.46±0.95 days, respectively.

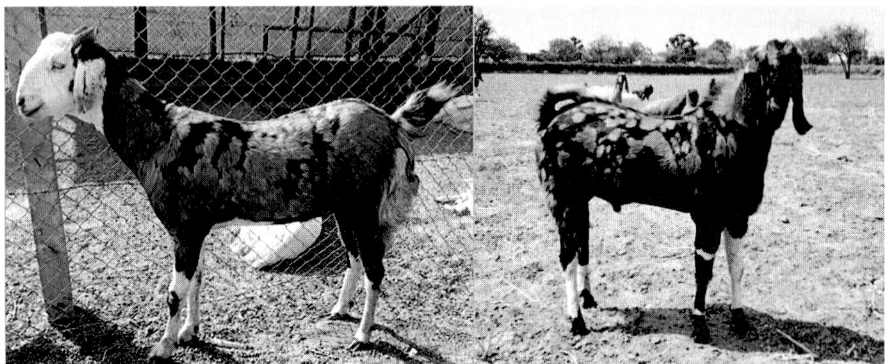

Fig. 21: Gujari goat breed: doe and buck

6. Sirohi

- **Sirohi is a breed from semi-arid region basically from Sirohi district Rajasthan**. However, this breed can be found in every part of India, as it is migrated by Goat Breeders from Rajasthan to other parts of the country.
- Synonyms: Majithi, Parbatsari, Devgarhi.
- It is **brown in colour with white patches**. Body size is compact and medium.
- The tail is twisted and carries coarse pointed short hair. Ears are long and flabby.
- Both male and female goats have horns, which are small, pointed, curved upward and backward.
- Buck weight is around 60-70 kg and Doe 30-35 kg. Birth weight is 2.5 to 2.75 kg. Kidding once in a year. Twinning common and average age at first kidding is 19 months.
- Quality of meat is good. Milk yield is 70 to 80 kg with 175 to 180 days lactation period.
- Population size is 19,52,116 as per 20th Livestock Census (2019).

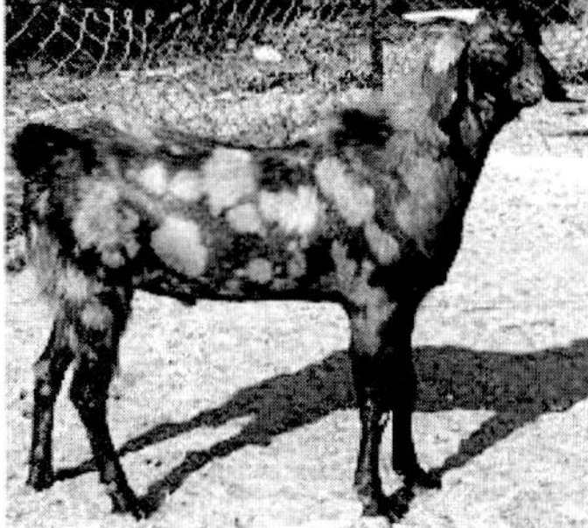

Fig. 22: Sirohi goat breed: doe and buck

7. Gohilwadi

- The Gohilwadi goat is an Indian goat breed which is distributed mainly through **Porbandar, Rajkot, Amreli, Bhavnagar and Junagarh districts of Gujarat state.**
- It is a multi-purpose goat breed, and generally raised **for meat, milk and also for fibre production.**
- The Gohilwadi goats are medium to large sized animals. They appear **mainly in black colour.**
- Their body is covered with coarse long hairs. Their nose line is slightly convex, and they have tubular and drooping ears.
- **Both bucks and does are generally horned. Their horns are slightly twisted, and turned backward.**
- Average body height of the mature doe is around 79 cm at the withers, and the buck's height is around 81 cm.
- Average live body weight of the mature bucks is around or up to 37 kg. And the mature does on an average can reach around or up to 36 kg live body weight.
- Population size is 2,88,453 as per 20th Livestock Census (2019).

Fig. 23: Gohilwadi goat breed: doe and buck

8. Kutchi

- The **Kutchi or Kathiawari**, is an important dual-purpose (meat and milk) goat breed, native to the Kutch district of Gujarat. They are medium-sized animals.
- Breeding tract: Patan, Kachchh, Mehsana and Banas Kantha district of Gujarat.
- According to the 18th Livestock Census 2007, the number of Kutchi goats in the country was 661,496.
- Both sexes have short, thick horns pointed upward.
- The coat is predominantly black, but a few white spotted animals are also found. Ears are medium in size, floppy and drooping with typical white markings.
- The coat is shaggy and dull in appearance with medium to long coarse hair. The annual yield of hair is about 200 g when shorn twice a year.
- Average milk yield is around 2 kg/day under stall fed conditions and 0.5 to 1.0 kg on grazing resources. The lactation length is about 6 to 7 months.
- Generally, there is one kidding annually with a twinning of 11%, which increases with supplementary feeding under stall fed conditions.
- Population size is 5,84,538 as per 20th Livestock Census (2019).

Fig. 24: Kutchi goat breed: doe and buck

9. Mehsana

- The Mehsana goat is a dual-purpose breed of domestic goat which is raised primarily for meat and milk production. The breed originates from the Gujarat state of India and is an important breed in its place of origin. Not surprisingly, the Mehsana goat breed derives its name from the origin place called "Mehsana" in Gujarat. The breed is found mostly in pure form in this area.
- Mehsana goat is a dual-purpose breed of domestic goats. It is raised for both milk and meat production. But the breed is also pretty good for hair production.
- The Mehsana goats are medium to large sized animals with a convex face profile. Their coat colour is **mainly black with white spots at the base of the ears**. The hairs on the coat are long and shaggy. Their nose line is straight.
- Ears of the Mehsana goats are of medium size and white in colour, which are leaf-like and dropping. Both bucks and does are bearded and usually have horns.
- Ears have white spots ranging from a few spots to complete white with few black spots at the base. White spots are present on the upper part of the upper muzzle and look like a ring in some of the animals. Hair coat is long and shaggy.
- Their horns have one or two twists and are curved upward and backward with pointed tips. Their tail is usually short and remains upward. The Udder of the Mehsana does is well developed, and the teats are large and conical.

- The bucks on average weigh about 37 kg, and average body weight of the does is about 32 kg.
- The Mehsana goats are pretty good milk producers and they produce about 1.3 kg of milk daily. Their first kidding age varies from 600 to 650 days. Single kidding is common, and twinning is low and ranges from 10-15%.

Fig. 25: Mehsana goat breed: doe and buck

- Population size is 4,22,509 as per 20th Livestock Census (2019).

10. Surti

- Name of the breed derives from the place called 'Surat' in Gujarat state of India. The breed is found in pure form in this area. But the total population of this breed is very small when compared to other goat breeds.
- Synonyms: Khandeshi, Kungi, Nimari, Desi.
- The Surti goat is distributed in the surrounding areas of Surat, Baroda of Gujarat and Nasik of Maharashtra.
- Surti goats are small to medium sized with a compact body. Their coat is predominantly white in colour with short and lustrous hairs. They have medium-sized dropping ears. Their forehead is prominent and the face profile is slightly raised.
- Both bucks and does usually have medium sized horns. Their horns are directed upward and backward. The udder of the does is well developed with large conical teats.
- They have relatively short legs, and they are usually unable to walk long distances.

- The does of this breed are comparatively larger than the bucks. The average weight of does and bucks is about 32 kg, and 30 kg, respectively.
- They are excellent milk producers, and produce about 2.0 to 2.25 kg of milk daily under a stall-fed system with supplementary feeding. But under village conditions with a grazing goat farming system, they produce about 1.2 kg of milk daily.
- The Surti goat is noted for its kidding percentage. The flock can be large within a very short period of time, because it can give birth to twins and triplets. Usually, the first kidding age varies from 400 to 500 days.
- Population size is 2,31,194 as per 20th Livestock Census (2019).

Fig. 26: Surti goat breed: doe and buck

11. Zalawadi

- It is believed that the Zalawadi goat breed originated in the former "Zalawad" region, which is now known as Surendranagar district and part of Zalawad falling in Rajkot districts **of Gujarat state in India.**
- **Synonyms: Tara Bakari**
- This **is one of the major goat breeds in Gujarat, and it is also known by its local name Tara Bakari.** They are usually reared by traditional shepherd community known as **Rabaris and Bharwads locally, in the semi-arid area of Saurashtra region**
- This is a multi-purpose breed of domestic goat because it is used for milk, meat and fibre production.
- They are medium to large-sized, **have long legs and erect corkscrew shaped horns**. Both bucks and does have twisted corkscrew shaped horns directed upward and backward with pointed tips.

- Their **coat colour is predominantly black** with lustrous, long shining hairs. Some goats also have black and white mixed coats. **Their face is slightly raised, ears are long, leafy and drooping and are invariably white speckled.**
- Their ears are so wide, loose and leafy that the shepherds traditionally trim the ears and sometimes split the ears to avoid injuries from thorny bushes while browsing.
- The Zalawadi does have a large udder and is well developed with long conical teats. Average body weight of the adult Zalawadi bucks and does is about 39 kg, and 33 kg. The Zalawadi does have a large udder and is well developed with long conical teats. Average body weight of the adult Zalawadi bucks and does is about 39 kg, and 33 kg. They are very beautiful animals and usually are of good behaviour.
- They have very good mothering ability and, on an average, a doe produces about 1.5 to 2.0 kg of milk daily. The birth rate is 55% twins and 2% triplets.
- Population size is 4,08,450 as per 20th Livestock Census (2019).

Fig. 27: Zalawadi goat breed: doe and buck

12. Kahmi

- This goat is native to **Saurashtra region of Gujarat**.
- Coat colour is unique, the neck and face are reddish brown while the rear abdominal part is black, locally called Kahmi.
- Ears are long, tubular and coiled, locally called veludi.
- Wattles are present in the majority of goats. Forehead is convex.
- Horns are directed upwards and backwards.

- These goats are used for both purposes i.e., meat and milk. Average daily milk yield is about 1.7 kg. Adult body weight is 56 kg in males and 48 kg in females. Average litter size is 1.4.

Fig. 28: Kahmi goat breed: doe and buck

13. Osmanabadi

- The breed derives its name from its habitat and is distributed in **Ahmednagar, Solapur and Osmanabad districts in Maharashtra**, where it is found in purest form.
- The body colour is mostly black but some animals with white patches on ears, neck and sometimes on body are also found. The hair coat is short and shining, face profile is straight, ears are of medium size and drooping.
- Five types of animals are available: 1. Entirely black with horns. 2. Entirely black with white ears and horns. 3. Entirely black and polled. 4. Entirely black, white ears and polled. 5. Brown and white patches from the face to lower side of the body
- Most of the males are horned but about 5% of the females are polled. Horns are straight/ curved and small in size (about 13 cm) running backward, upward and downward.
- Wattles are found in a number of animals. On an average, adult males and females weigh 34 kg and 32 kg, respectively.
- Osmanabadi goats mostly have black coat colour, with grey white skin, black muzzle, eyelids and hooves, possessing grey straight and backward oriented horns (13.01 cm length), pendulous ears (4.83±0.08 cm length), convex head type, absence of wattles and beard, and curved, slender and medium-length tail Osmanabadi is a medium sized breed with comparatively long body and long legs.

- The average body length, body height and heart girth are 69, 77 and 72 cm in males and 68, 75 and 72 cm in females, respectively.
- Performance: The average age at first kidding and kidding interval is 523 and 214 days respectively. Goats of this breed have very efficient reproduction and in well managed flocks, with 30% twining and 2% triplets.
- The daily milk yield ranged from 700 g to 1500 g under well managed village flocks with a lactation length of 130-150 days.
- The observed overall survivability rate of Osmanbadi goat under farmers' condition as low as 68.87%, whereas in adult and kids it was found to be 81.60% and 47.16%, respectively.
- Population size is 35,97,071 as per 20th Livestock Census (2019). This breed shares 2.4% and fourth among total population and ranks fourth.

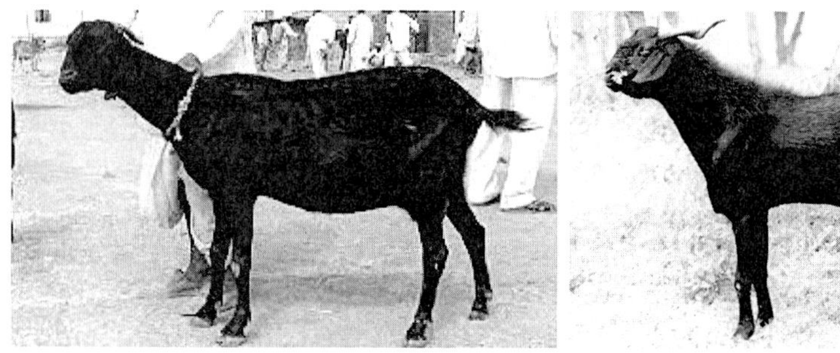

Fig. 29: Osmanabadi goat breed: doe and buck

14. Sangamneri

- The semi-arid region of Maharashtra comprising of **Nasik, Ahmednagar and Pune** districts forms the native habitat of the Sangamneri goat breed.
- The breed derives its name from the **Sangamner Tehsil of Ahmednagar district**. They are medium-sized goats.
- The coat is completely **white with mixtures of black and brown. Ears are long and drooping.**
- Both sexes have horns directed backward and upward.
- Horns are thin pointed, directed backward and upward. Average length is 12.71cm in females and 19.44 cm in males.

- Hair coat is extensively coarse and short.
- The litter size is mainly single. However, 15 – 20% goats produced twins but triplets are rare.
- The average daily milk yield varies between 0.5 to 1.0 kg with an average lactation length of about 160 days.
- Although this breed is reared mainly for meat, some goats show a good milch potential.
- Dressing percentage is about 41% at 6 months, 45% at 9 months and 46% at 12 months of age.
- According to the 18th Livestock Census (2007), the numbers of Sangamneri goats in the country are 15,000 to 20,000 (NBAGR). In 2013, estimated population size was 1,62,834.
- Population size is 1,63,091 as per 20th Livestock Census (2019).

Fig. 30: Sangamneri goat breed: doe and buck

15. Konkan Kanyal

- Konkan Kanyal goat population belonging to different sex and age groups existing in the different villages of **Kudal, Dodamargi, Malvan, Verungla and Sawantwari tehsils of Sindhudurg district of Konkan region of Maharashtra** state of India.
- They are mainly reared under free range systems. Goats are **predominantly black with white markings in a specific pattern**, however, brown coats with the similar pattern of white or light brown markings are also seen.
- Colour: Black with white marking on collar, lower jaw and ventral surface

- Horn shape and size: Cylindrical, backward and medium in size (15-25 cm)
- Visible characteristics: Bilateral white strips from nostrils to ear. Legs are long, laterally black, medially white and white from knee to fetlock joint. Tail is dorsally black and ventrally white.
- The body weights of adult males and females were 35.26 kg and 29.80 kg, respectively under field conditions.
- The average milk production per lactation was 69.035 litre in a lactation period of 96.199±17.49 days.
- The meat of Kanyal goats is red with fine and firm fibre.
- These goats are raised under the closed housing with the floor.
- Mortality reported was about 3.0%, hence these goats are well adapted to the agro-ecological conditions of Konkan region.
- Population size is 16,892 as per 20th Livestock Census (2019).

Fig. 31: Konkan Kanyal goat breed: doe and buck

16. Berari

- The Berari goat, a breed of central region of India (**Vidarbha region of Maharashtra**), is a low yielding **prolific meat breed** thriving well in tropical wet and dry climatic conditions.
- Synonyms: Lakhi, Gaorani
- Origin: Originated from Berar region of Central Province and Berar, which is recently known as Vidarbha region of Maharashtra state.
- The average daily milk yield of Berari goats is 533.28 gm/day with a total lactation yield of 78 kg in 130 days of lactation, while age at first conception is 10 to 11 months.

- This breed is having a good prolificacy i.e., 57% twinning, 2% triplets/quadruplets.
- **The coat colour of the Berari goat ranges from dark tan to light tan**. Horns were noticed in both the sexes and were straight, orientation of horn was mostly found as **upward-backward**. **Black hair line along the vertebral column extending up to the tail is also seen.**
- All the Berari goats had a pendulous (drooping), flat and leafy ear and convex forehead. Majority of these goats (98%) do not have beard and wattle and does have bowl shaped udder (89%) with conical teat shape and pointed teats tip.
- The mean chest girth, body length, height at wither, paunch girth, horn length, ear length and tail length for adult animals (above 2 years of age) were 73.3, 63.1, 71.3, 76.2, 9.6, 15.7 and 13.7 cm, respectively.
- The body weights in males and females at 1-2 years of age were 27.4 and 24.1 kg and that of adult goats 34.5 and 27.3 kg, respectively.
- Light to dark strip on lateral sides from base of horn to nostrils, black colour ring around neck in adult males, and black hair line along with the vertebral column extending up to tail in both sexes was observed as a unique characteristic in Berari goat.
- Population size is 84,823 as per 20th Livestock Census (2019).

Fig. 32: Berari goat breed: doe and buck

17. Anjori Goat

- It is a medium sized, meat purpose breed.
- It is distributed in **Raipur, Durg, Rajnandgaon, Kanker, Dhamtari, Mahasamund districts of Chhattisgarh state**.
- Majority of animals are brown in colour.
- It is hardy and well adapted to the local ecosystem.

- Average adult body weight of males and females is 35 kg and 28 kg, respectively.
- Average milk production per lactation is 26 kg.

Fig. 33: Anjori goat breed: doe and buck

IV. Southern region

The states under this southern region include parts of Karnataka, Andhra Pradesh, Tamil Nadu and Kerala states.

1. Bidri

- **Origin Local:** Named after its place of origin i.e., Bidar. Bidri is a medium-sized black goat **with higher fecundity and is distributed in Bidar and Kalaburagi district of Karnataka** and adjoining areas.
- Bidar plateau has an elevation range from 640 to 684 m above mean sea level. The ground surface is flat, gently sloping forming broad valleys and flat-topped hills. The main occupation of the people in the district is agriculture and related operations. The main food crops are Jowar, Paddy, Wheat, Bajra, Maize and pulses. Groundnut, sugarcane and cotton are the cash crops.
- This breed has higher fecundity and twinning is very common.
- **Breeding tract:** Native tract is bounded by Nizamabad and Medak districts of **Telangana state on the eastern side, Latur and Osmanabad districts of Maharashtra state on the western side, Nanded district of Maharashtra state on the northern side.**
- **Adaptability to environment: They are** well adapted to the agro-ecological conditions of the region and sustained mainly on grazing management.

- Management: Average flock size is 74.3 (21-130). Goats were housed in kutcha open sheds during night only.
- Colour: Black, some have white spots on ears, forehead, neck and knees
- Horn shape and size: Curved and directed backward, outward and downward. Females 13.72 cm, Males 16.44cm.
- Forehead is straight. Muzzle, eyelids and hooves are black. Ears are pendulous. Udder is hairy and small in size.
- Goats are reared for meat only. Milking is not practiced. Twining is common but first kidding single.
- Bidri is a meat type goat with 50% dressing percentage.

	Male	Female
Height (avg. cm.)	79.25	74.84
Body Length (avg. cm.)	58.17	56.09
Heart girth (avg. cm.)	80.75	77.12
Body weight (avg. kg.)	36.78	32.36
Birth weight (avg. kg.)	2.5	2.2

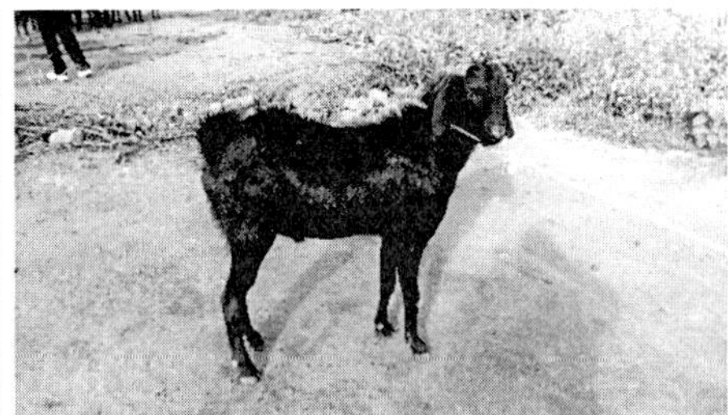

Fig. 34: Bidri goat breed: doe and buck

2. Nandidurga

- Synonyms: Nandi, Bilimeke
- Origin: Named after its place of origin i.e., Chitradurga district of Karnataka and white colour.
- Used for meat purposes only. This breed has higher fecundity and twinning is very common.
- **Breeding tract:** Chitradurga district lies in the valley of the Vedavati River, with the Tungabhadra River flowing in the northwest. Average elevation of the tract ranges from 550 to 822 meters. These are distributed

in Chitradurga, Tumkur and Davangere districts of Karnataka. Tract is bounded by Anantpur district of Andhra Pradesh on the eastern side, Shivmoga, Chikkamagaluru and Hassan districts of Karnataka on the western side.

- Different types of soils, climatic factors and irrigation facilities have created favourable conditions for extensive cultivation of horticulture crops like Coconut, Arecanut, Banana, etc. Kharif is the main cropping season with Paddy, Ragi, Jowar and Maize crops. Considerable amounts of mineral wealth and natural resources such as iron ore, quartz, silver sand, graphite, corundum, dolomite, clay and soap stone are also found.
- Adaptability: Nandidurga goats are adapted to hard rocky areas and graze efficiently on hillocks. Management system is extensive.
- Comments on Management: Separate sheds are provided for housing of goats during night only. Majority of houses are kutcha. Goats are raised mainly on grazing. Leaves of Acacia, Neem (*Azadirachta indica*) and Tamarind (*Tamarindus indica*) are also fed. Kids are fed with maize and groundnut cake powder. Milking is not practiced.
- **Colour:** White, some have black/brown spots on ears, forehead, neck and knees.
- **Horn shape and size:** Curved, directed backward, downward, inward and touching neck in few. Average size: male 20.33cm, female 15.79.
- **Visible characteristics:** Muzzle, eyelids and hooves are black. Ears are leafy and pendulous. Udder is hairy and pendulous. Teats are funnel shaped.
- Average age at slaughter varies from 8 to 10 months and dressing percentage is around 50%. Males mature earlier than females.
- Averages of some body parameters are given below:

	Male	**Female**
Height (cm.)	79.4	71.92
Body length (cm.)	59.65	55.31
Heart girth (cm.)	81.47	75.07
Body weight (kg.)	38.92	30.11
Birth weight (kg.)	3.32	2.28

Fig. 35: Nandidurga goat breed: doe and buck

3. Attapady Black

- Breeding tract: Mainly found in Attapady block, Palakkad or Palghat in Kerala. The tract consists of high range areas of Kerala between Nilgiri hills in the north and the Muthikulam hills in south.
- This is a mutton breed of Kerala.
- Adaptability: Attapady black goats are well adapted to the agro-ecological conditions of the Attapady region and maintained mainly on grazing.
- These goats have better immunity than the goats of other breeds.
- Comments on Management: Flock size ranges from 5 to 80. Housed in the night only.
- **Colour: Their** colour is black and the extremities including face, ears and legs are also black. White spots on the forehead are noticed in some animals. The colour of skin is light black and sometimes grey.
- **Horn shape and size:** Curved and oriented backwards. Small in size.
- Visible characteristics: Extremities are black. Tail is a bunchy type.
- Current population size is only 31,182 (20th Livestock Census, 2019).
- Averages of some body parameters are given below:

	Male	**Female**
Height (cm.)	79.5	66.8
Body Length (cm.)	66.6	62.8
Heart girth (cm.)	71.0	68.9
Body weight (kg.)	34.5	31.3
Birth weight (kg.)	1.73	1.6

Fig. 36: Attapady Black goat breed: doe and buck

4. Malabari

- Synonyms: Thalassery, Tellicherry, Cutch, West coast, Malabar
- Origin: Name derived from Malabar region of Kerala, found in Calicut Kannur / Cannanore, Wayanad, Malappuram in Kerala.
- Major utility: Milk and Meat purposes.
- Adaptability to environment: It is adapted to the hot and humid conditions of Kerala.
- **Colour:** Majority complete white. Some are black or brown or admixtures.
- **Horn shape and size:** Slightly twisted horns directed outward and upward. In some cases, curved backward and downward touching the skin. Small in size.
- Medium-sized ears, directed outward and downward reaching up to the nose. Presence of wattles from the neck is a visible characteristic of this breed.
- **Malabari goat is a highly prolific goat and litter size varies from 2 to 4. Malabari bucks have been used widely for upgrading nondescript goats.**
- Averages of some body parameters are given below:

	Male	**Female**
Height (cm.)	76.29	64.46
Body Length (cm.)	77.29	68.38
Heart girth (cm.)	79.17	73.0
Body weight (kg.)	38.96	31.12
Birth weight (kg.)	2.22	2.0

Fig. 37: Malabari goat breed: doe and buck

5. Kodi Adu

Synonyms: Porai Adu

Origin: The name Kodi Adu means goat with long and lean body conformation and its synonym Porai Adu has been derived from the body colour viz., splashes of black or reddish-brown colour over the white coat.

Breeding tract: The major part of the breeding tract lies in the southern agro-climatic zone of Thoothukudi/Tuticorin, Ramanathapuram in Tamil Nadu and the climate is generally hot, semi-arid and tropical in nature. The mean annual maximum and minimum temperatures are 32.9 and 24.5 °C, respectively. The mean annual relative humidity recorded at 08.30 h and 17.30 h is 78.0 and 73.6%, respectively. The average annual rainfall is 668.0 mm, received in 35.4 rainy days.

Utility: Kodi Adu goats are reared for meat, skin and manure. They serve as the sole or subsidiary source of livelihood for a large number of small and marginal farmers as well as for landless labourers in the breeding tract.

- Adaptation to cover long distances during browsing.

Management: They are allowed to browse extensively as a herded group for a distance of 5 to 10 km. Goats are housed only during nights. Few male kids are selected based on body weight for breeding purposes and others are disposed off and almost all the females are retained. The goat owners mainly belong to Yadhava, Devar and Maravar communities.

Colour: White with splashes of black or reddish-brown colour

Horn shape and size: Directed upward, backward and curved downward or upward and sharp at the tip. Size 15-25 cm.

Characteristics: Kodi Adu goats are tall, long, lean and leggy animals with compact bodies. Females have short, straight and sleek hairs on almost all parts of the body. Males have fairly long, straight and rough hairs on the neck and withers.

- Averages of some body parameters are given below:

	Male	**Female**
Height (. cm.)	82.39	79.16
Body Length (cm.)	74.08	71.94
Heart girth (cm.)	75.46	73.79
Body weight (kg.)	33.43	30.94
Birth weight (kg.)	2.91	2.76

Fig. 38: Kodi Adu goat breed: doe and buck

6. Salem Black

Synonyms: Karuppadu (in vernacular Tamil language)

Origin is indigenous: The name Salem Black is derived based on the body colour (i.e., complete black body colour) and place of origin i.e., Salem district of Tamil Nadu. They are found at Krishnagiri, Erode / Periyar, Dharmapuri in Tamil Nadu.

Breeding tract: They have a confined breeding tract in north-western agro-climatic region of Tamil Nadu, viz., Dharmapuri and Karimangalam blocks of Dharmapuri taluk, Pennagaram block of Pennagaram taluk and Palacode taluk of Dharmapuri district, Mecheri and Kolathur blocks of Mettur taluk of Salem District, Thalavadi block and Andiyur block of Erode district. The mean annual maximum and minimum temperatures are 34.3°C and 21.9°C, respectively. The mean annual relative humidity recorded at 08.30 h and 17.30 h are 77.2 and 55.1%, respectively. The average annual rainfall is 1112 mm.

- **Utility:** They are reared for meat, skin and manure purposes. Meat of this breed is considered to be very tasty compared to other goats and local preference is totally for chevon from this breed. Does are not milked.
- Well adapted to the harsh climatic conditions (hot, semi-arid and tropical) of North-western parts of Tamil Nadu.
- **Management:** Goats are taken out for browsing up to a distance of three to six kilometres for a period of seven to eight hours daily in the bushes along road sides, uncultivable lands, forest areas and harvested fields. During summer months due to the scarcity of fodder the goats are fed with Acacia pods, tapioca leaves and rind, and dry fodders like groundnut hay, paddy straw and horse gram hay. Thrive well in poor fodder resource areas and utilize agri by-products (groundnut haulms and husk and gingili (sesame) haulms etc.) Feeding of animals with concentrates is also practiced in some flocks. Most of the management activities are carried out by women. Housed mostly during nights in the sheds or pens in the agricultural fields.

Colour: Completely black with glossy hair coat.

Horn shape and size: No typical horn pattern however mostly directed upward and backward and sharp at the tip. Males: 20 cm, Females: 13.7 cm.

Visible characteristics: Salem Black goats are tall, long, lean and leggy animals with compact bodies. Head is medium in length. The eyes are small and bright and the eyelashes are black in colour. The ears are medium in length, leaf-like and semi-pendulous. Neck is thick, broad and well set to the thorax in males.

- Averages of some body parameters are given below:

	Male	**Female**
Height (avg. cm.)	80.1	73.6
Body Length (avg. cm.)	70.2	67.1
Heart girth (avg. cm.)	73.7	69.1
Body weight (avg. kg.)	38.16	31.58
Birth weight (avg. kg.)	2.27	2.22

Fig. 39: Salem Black goat breed: doe and buck

7. Kanni Adu

Synonyms: Varikan Adu, Karapu Adu, Pullai Adu

Origin: The names Kanni Adu and Varikan Adu are derived from the presence of stripes on either side of the face.

Utility: They are reared for meat, skin and manure purposes.

Breeding tract: Sattur and Sivakashi taluks of Virudnagar district; Kovilpatti and Vilattikulam taluks of Tuticorin district; Sankarankovil taluk of Tirunelveli district of Tamilnadu.

Management: Goats generally maintained on grazing for which they have to travel 5 to 10 km. **Colour:** Generally, black with two white stripes on either side of the face and white colour in the underbelly and inner side of the legs.

Horn shape and size: Directed backward and outward; backward outward and curved upward are predominantly present. Medium sized (15-25 cm) in males and small (<15 cm) in females.

Characteristics: These goats have white strips on either side of the face extending from the base of the horn to the corner of the muzzle. White lines on the edge of ears. Majority of the animals have white patch or line on either side of the neck.

Averages of some body parameters are given below:

	Male	Female
Height (cm.)	81.46	74.83
Body Length (cm.)	73.06	68.43
Heart girth (cm.)	75.22	70.07
Body weight (kg.)	34.05	28.17
Birth weight (kg.)	2.11	2.05

Fig. 40: Kanni Adu goat breed: doe and buck

V. East region breeds

1. Teressa

- Teressa goat is the first indigenous goat breed from **Andaman & Nicobar Islands** is registered at ICAR- National Bureau of Animal Genetic Resources, Karnal, (Haryana).
- In Nicobari language it is known as Pookore which means goat.
- **Due to their origin and distribution in Teressa Islands and other islands of Nicobar it is popularly known as Teressa goat.**
- Native breeding tract is mainly found in the Nicobar group of islands viz. Teressa Island, Car Nicobar, Katchal, Nancowrie etc.
- However, a sparse population is also available at Little Andaman and Andaman Islands also.
- They are generally tall, sturdy, brownish or dark tan or black or white in colour with white and black patches. They have black hairs on the dorsal midline which goes up to the tail.
- Peculiar white patch/line starting from inner canthus of both eyes or from eyebrows and extending up to nostrils or mouth. The muzzle, eyelids and hoofs are black.
- Height ranges from 24 to 27 inches. Head length of male and female is 8.86±0.10 and 7.5±0.27 inches, respectively.
- Tail is medium to long. The Horns is large with a flat at its base. Ears are erected and directed towards downwards. Age at sexual maturity is about 8-9 months. Body weight at 12 months varies from 20-25 kg, weight at 2 years varies from 35-40 kg, and at 4 years varies from 50-65 kg.

- Average milk yield may go up to 1 litre/day. Twinning is common and it is more than 53%.
- The present population of this breed is less than 9000, which needs attention for conservation and further propagation in the native breeding tract.
- Benefit: This breed is highly suitable and adaptable to the island's climatic conditions.
- Population size is 3,362 as per 20th Livestock Census (2019).

Fig. 41: Teressa goat breed: doe and buck

2. Assam Hill

- Distributed in **Assam and adjoining areas of Meghalaya**. Assam Hill goats are mostly white with occasional black patches on the backline and legs.
- **Synonyms**: Asomi.
- **Breeding tract**: Some animals are also available in Sonitpur, Sivsagar, Tinsukia, Kamrup, Jorhat, Hailakandi, Goalpara, Darrang, Lakhimpur, kokrajhar, Karimganj, Bongaigaon, Borpeta, Baksa, Nalbari and Udalguri districts of Assam.
- These goats are short legged with small body size.
- Both buck and does are bearded and have short cylindrical horns, which are directed upwards and outwards.
- **Colour**: Usually white with occasional black patches on backline and legs.
- **Horn shape and size**: Cylindrical and tapering towards the end (corrugated) and pointed at the tip. Usually straight but in some animals slightly backward. Small in size.

- These goats are short legged with small body size. Both buck and does are bearded.
- Ears are medium in size, horizontally placed with pointed tips. Tail is short and hairy.
- This is an important meat type animal with high prolificacy.
- These goats are reared **mainly for meat**. Adult body weight ranges from 15 to 26 kg.
- Age at first kidding ranges from 337 to 447 days. Average litter size is 1.6.

Fig. 42: Assam Hill goat breed: doe and buck

3. Sumi-Ne

- **Sumi-Ne goats are mainly found in Zunehoboto and Tuensang districts, whereas few goats are also found in Kiphire, Phek and other districts of Nagaland.**
- **Synonyms: Apu-Asu-Ne, Nagaland Long Haired.**
- The long hair goat, as the name indicates, is distinguished from other goat populations of NEH region by the presence of long silky hair in males and are reared by Sumi tribe people under extensive and semi-extensive system of management.
- These goats are predominantly of black (head & neck) and white (remaining parts) colour.
- These goats have long hairs, which are used for commercial purposes by the tribal.
- They have a straight head, horizontal ears, **pointed horns and a beard is present in all goats.** Presence of long hair in adult animals is the most important phenotypic characteristic. The length of the fibre, however, is more in case of male as compared to that of females.

- Horns shape and size: small sized and slightly curved backward.
- Colour: white with characteristic black markings on head, neck and legs.
- Population size is 1,509 as per 20th Livestock Census (2019).

Table: Averages of some body parameters Sumi-Ne goats

Traits/ Parameters	Male	Female
Height (cm)	46.45	45.8
Body length (cm	50.5	45.53
Heart girth (cm)	52.4	47.5
Weight (kg)	16.18	13.5
Birth weight (kg)	1.3	1.22

Fig. 43: Sumi-Ne goat breed: doe and buck

4. Ganjam

- The Ganjam breed, also known as Dalua, is found in eastern India, primarily in the **Gajapati, Rayagada and Koraput districts of Odisha.**
- Synonyms: Dalua, Baigani, Gola goat, Lanka goat.
- A small number are also found in Sikkim. They are tall, leggy animals. The coat may be black, white, brown or spotted, but black predominates.
- The hairs are short and lustrous. Ears are medium sized and both bucks and does have long, straight horns, directed upward with a medium-length tail.
- **Colour:** Black or brown black. White, brown and spotted animals are also found.
- **Horn shape and size:** Twisted and curved. Long (up to 50 cm in many cases), parallel and pointed backward and upward.
- Males usually have beards. They have convex heads, long and drooping ears, and wattles present in both sexes.

- The kidding percentage is 82 and the litter size is primarily single (98.4%). Kidding takes place once a year.
- Milk yield is about 65 kg in an average lactation period of 150 days.
- According to the 18th Livestock Census 2007, the number of Ganjam goats in the country was 148,473. However, as per 20th Livestock Census (2019) the population size is 2,11,478.

Fig. 44: Ganjam goat breed: doe and buck

5. **Black Bengal**

- The Black Bengal is found in the **eastern region of India, in the states of West Bengal and adjoining areas in Jharkhand, Bihar and Orissa, Assam, Mizoram, Tripura and** a few numbers also found in Jammu and Kashmir, Himachal Pradesh and Punjab.
- This is a very popular breed of goat in Bangladesh because of its low demand for feed and high kid production rate.
- **Coat colour is predominantly black, brown, grey and white with soft, glossy and short hairs.**
- **Horn shape and size are small (<15), directed upward and sometimes backward.**
- **These are small-legged goats. Hair coat is short and lustrous.**
- **Nose line is slightly depressed.**
- Dwarf in body size, legs short, straight back; both sexes are bearded.
- An adult male goat weighs about 25 to 30 kg and female 20 to 25 kg. It is poor in milk production.
- **Most prolific among the Indian breeds**. Multiple births are common - two, three or four kids are born at a time. Kidding is twice a year.
- The Black Bengal goats, both male and female gain sexual maturity at an earlier age i.e. at around 6 months of age, earlier than most other breeds.

- Average litter size is 2.1. Average age at first kidding 9-10 months. Average lactation yield is 53 kg. Lactation length is 90 to 120 days.
- **Their skin is of great demand for high quality shoe-making.** The meat is excellent and palatable.
- Milk yield is low and is barely sufficient to feed the kids. Skin of the Black Bengal goat is used for making chamois leather a highly valued specialty leather.
- Population size is 2,76,61,976 as per 20th Livestock Census (2019), which is the highest in all goat breeds of India. This breed shares 18.6 % among all the Indian goats.

Fig. 45: Black Bengal goat breed: doe and buck

6. Andamani

- Andamani goat is a medium sized, meat purpose breed, mainly distributed in the Middle and North Andaman Islands.
- It is well adapted to the tropical humid climate of the Island and is preferred for the excellent chevon quality.
- Average adult body weight for males is 29 kg and for females is 26 kg. Average milk production per lactation is 29 kg.

Fig. 46: Andamani goat breed: doe and buck

Table 6: Important characteristics of some breeds

Sl. No.	Name of the Breed	Native Tract	Peculiar Characteristics
1	Andamani	Middle and North Andaman Island	It is well adapted to the tropical humid climate of the Island and is preferred for the excellent chevon quality.
2	Anjori	Raipur, Durg, Rajnandgaon, Kanker, Dhamtari, Mahasamund districts of Chhattisgarh state.	Majority of animals are brown in colour. It is hardy and well adapted to the local ecosystem.
3	Assam Hill	Assam and Meghalaya	Mostly white with occasional black patches on backline and legs. Short legged with small body size.
4	Attapady	Kerala	Black in colour with bronze-coloured eyes.
5	Barbari *(city breed)*	UP and Rajasthan	Dwarf breed, highly suitable for stall feeding.
6	Beetal	Punjab	Roman nose, buck has beard, females are beardless.
7	Berari	Maharashtra	Black coloured ring around neck in adult male. Black hair line along with the vertebral column extending up to tail.
8	Bhakarwali	Jammu and Kashmir	Whole body is covered with long hairs. Ears are cut and pendulous, Horns are screw type.
9	Bidri	Karnataka	Medium sized black goat with higher fecundity.
10	Black Bengal	West Bengal	Best chevon breed of India. Highly prolific goat breed.
11	Changthangi	Jammu and Kashmir	***'Pashmina'*** producing breed. Pashmina used for making high quality Kashmiri "***rug or shawl***".
12	Chegu	Himachal Pradesh	***'Pashmina'*** producing breed.
13	Gaddi	Himachal Pradesh	Prominent and alert eyes and roman nose.
14	Ganjam	Orissa	Brown, black and grey coat colour (rarely- white)
15	Gohilwadi	Gujarat	Body is covered with coarse long hairs, black body coat.
16	Gurari	Jaipur and Sikar districts of Rajasthan.	The animals are brown and white mixed coat colour, while white coloured face, leg and abdomen are typical features of the breed.
17	Jakhrana	Rajasthan	Skin popular for tanning purpose.
18	Jamunapari	Uttar Pradesh	Biggest and most majestic breed of Indian goat. *Parrot mouth* appearance. Most beautiful breed.
19	Kahmi	Gujarat	Neck and face are reddish brown while rear abdominal part is black.

Sl. No.	Name of the Breed	Native Tract	Peculiar Characteristics
20	Kanni Adu	Tamil Nadu	Black or white spots in the black background.
21	Kodi Adu	Tamil Nadu	Taller and found with different colours, but predominantly black
22	Konkan Kanyal	Maharashtra	Predominantly black with white marking in a specific pattern.
23	Karauli	Sawai Madhopur, Kota, Bundi, and Baran districts of Rajasthan.	Ears in Karauli goats are long, pendulous with folded and brown lines on the border of ears. Karauli bucks have prominent hanging dewlap.
24	Kutchi	Gujarat	Corkscrew shaped horns.
25	Malabari (Tellicherry)	Kerala	Seen in white, purple and black colours. Mainly for meat purpose.
26	Marwari	Rajasthan	Less prone to diseases, hair quality is lustrous, short tail.
27	Mehsana	Gujarat	Coat colour is mainly black with white spots at the base of the ears, The hair on the coat is long and shaggy.
28	Nandidurga	Karnataka	White coloured and used mainly for meat purpose.
29	Osmanabadi	Maharashtra	Coat colour is predominantly black; white, brown and spotted occur.
30	Pantja	Uttarakhand and Uttar Pradesh	These goats are well adapted to humid condition of Tarai region. Twining is common in Pantja goat.
31	Rohilkhandi	Uttar Pradesh	Coat colour- mainly black with star or patch on neck and face
32	Salem Black	Tamil Nadu	Completely black in colour, mainly for meat purpose.
33	Sangamneri	Maharashtra	Coat colour is complete white but admixtures of black and brown are also available.
34	Sirohi	Rajasthan and Gujarat	Tail twisted and carries coarse pointed hair. Coat colour is brown, white, and admixture of colours in typical patches.
35	Sojat	Sojat, Pipar, Pali, Jodhpur, Nagaur and Jaisalmer districts of Rajasthan.	The coat colour of these animals is white with brown spots on head, neck, ear and legs; however, pure white animals are also found in the field.
36	Sumi-Ne	Nagaland	Coat colour is white with black patches on head, neck and legs. Also, known as **"Nagaland long hair goat".**
37	Surti	Gujarat	Best dairy goat in India. Mainly white colour body coat.
38	Teressa	Andaman & Nicobar	Tall, brownish or black or white in colour with white and black patches.
39	Zalawadi	Gujarat	Long legs and erect corkscrew shaped horns. Wide loose and leafy ears.

5

Poultry Breeds of India

Origin of Fowl

Domestication of poultry seems to have been undertaken in south-east Asia. The chickens were brought to India by 1000 BC, and later on they spread north, westwards and reached Greece by 525 BC. By the beginning of Christian era, the birds were already popular in West Asia and East Europe, and then gradually reached South Africa, Australia, Japan, USSR and USA.

Red Jungle fowls are the ancestors of the present-day poultry breeds. There are four known species of wild fowl and they belong to the same genus "Gallus" meaning a cock. The four species are as follows:

- *Gallus gallus* or *Gallus bankiva* - Red Jungle Fowl
- *Gallus lafayetti* - Ceylon Jungle Fowl
- *Gallus sonneratii* - Grey Jungle Fowl
- *Gallus varius* - Javan Jungle Fowl

The Red Jungle fowl is supposed to be the main contributor for the development of modern-day poultry and is widely distributed throughout Burma, China, India, Philippines, Sumatra and Thailand. The plumage (colour of the feather) of females resembles that of Brown Leghorn, the males have orange red feathers in hackle, wing bow and saddle regions, while the breast is black. Eggs are buff (off-white) in colour. The legs are slate (dark bluish grey) coloured. The comb is all red.

Fig. 1: Red Jungle fowl cocks (upper row) and hens (lower row)

There are five sub-species of Red Jungle fowl which are as follows:

- *Gallus gallus bankiva* - Javan Red Jungle Fowl
- *Gallus gallus gallus* - Cochin-Chinese Red Jungle Fowl
- *Gallus gallus jabouillei* - Tonkinese Red Jungle Fowl
- *Gallus gallus murghi* - Indian Red Jungle Fowl
- *Gallus gallus spadaceus* - Burmese Red Jungle Fowl

In addition to these, the related grey jungle fowl (*G. sonneratii*), Sri Lankan jungle fowl (*G. lafayettii*) and the Javanese green jungle fowl (*G. varius*) have also contributed genetic material to the gene pool of the modern chicken.

There are many species of birds included under poultry *viz.* chicken, duck, geese, quail, turkey, guinea fowl etc. and within each species there are many breeds available. Different breeds available under each species are discussed in this section.

Body Parts of a Chicken

Before learning about the different breeds, varieties and strains of chicken, it is suggested to go through the different parts of a chicken thoroughly for better understanding.

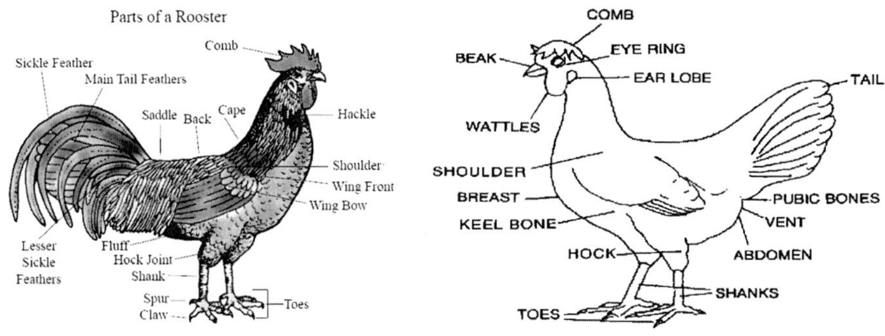

Fig. 2: Body Parts of a chicken

Zoological classification of common poultry species

Poultry refers to a wide variety of birds of several species and it refers to them whether they are alive or slaughtered, dressed and prepared for market.

Table 1: Zoological classification of family *Phasianidae*

1. Fowl	2. Japanese quail	3. Pheasant	4. Peafowl
Kingdom – Animalia	Kingdom – Animalia	Kingdom – Animalia	Kingdom – Animalia
Phylum – Chordata	Phylum – Chordata	Phylum – Chordata	Phylum – Chordata
Subphylum – Vertebra	Subphylum – Vertebra	Subphylum – Vertebra	Subphylum – Vertebra
Class – Aves	Class – Aves	Class – Aves	Class – Aves
Subclass – Neornithes	Subclass – Neornithes	Subclass – Neornithes	Subclass – Neornithes
Order – Galliforms	Order – Galliforms	Order – Galliforms	Order – Galliforms
Family – *Phasianidae*	Family – *Phasianidae*	Family – *Phasianidae*	Family – *Phasianidae*
Genus – *Gallus*	Genus – *Cotunix*	Genus – *Phasians*	Genus – *Pavo*
Species – *domesticus*	Species – *japonica*	Species – *colchricus*	Species – *cristatus*

Table 2: Zoological classification of family *Meleagridae* and *Antidae*

5. Turkey	6. Goose	7. Muscovy	8. Duck
Kindom – Animalia	Kindom – Animalia	Kindom – Animalia	Kindom – Animalia
Phylum – Chordata	Phylum – Chordata	Phylum – Chordata	Phylum – Chordata
Subphylum – Vertebra	Subphylum – Vertebra	Subphylum – Vertebra	Subphylum – Vertebra
Class – Aves	Class – Aves	Class – Aves	Class – Aves
Subclass – Neornithes	Subclass – Neornithes	Subclass – Neornithes	Subclass – Neornithes
Order – Galliforms	Order – Ansariforms	Order – Ansariforms	Order – Ansariforms

Family – *Meleagridae*	Family – *Antidae*	Family – *Antidae*	Family – *Antidae*
Genus – *Meleagris*	Genus – *Anser*	Genus – *Cairina*	Genus – *Anas*
Species – *gallopavo*	Species – *anser*	Species – *moschata*	Species – *platyrhynchos*

Table 3: Zoological classification of family *Struthionidae*

9. Ostrich
Kingdom – Animalia
Phylum: Chordata
Subphylum: Vertebrata
Class: Aves (wings and feathers)
Subclass – Neornithes
Order: Struthioniformes
Family: *Struthionidae*
Genus: *Struthio*
Species: *camelus*

General Information About Poultry

Table 4: Common terms used in poultry

Species	Young	Adult	Adult
		Male	Female
Chicken	Chick	Cock	Hen
Duck	Duckling	Drake	Duck
Turkey	Poult	Tom	Hen
Quail	Chick	Cock	Hen
Guinea fowl	Keet	Cock	Hen
Goose	Goosling	Gander	Goose
Pigeon	Squab	Pigeon	Pigeon
Swan	Cygnet	Swan	Swan

Table 5: Incubation period, Chromosome numbers and age at sexual maturity of different species of poultry

Sl.No	Species	Incubation period (days)	Chromosome number (pairs)	Age at sexual maturity (weeks)
1	Chicken	21	78 (39)	18-20
2	Duck	28	80 (40)	28-30
3	Muscovy duck	33-35	80 (40)	28-30
4	Goose	28-32	80 (40)	28-30
5	Guinea fowl	27-28	78 (39)	28-32
6	Turkey	28	80 (40)	28-30
7	Quail	17-18	78 (39)	06-07
8	Pigeon	18	78 (39)	10-12
9	Ostrich	42	80 (40)	52
10	Emu	52-55	80 (40)	52

Table 6: Classification of chicken

Class	American	Asiatic	English	Mediterranean
Shank	Clean	Feathered	Clean	Clean
Skin colour	Yellow	Yellow	White	Yellow or white
Earlobe colour	Red	Red	Red	White
Purpose	Dual	Meat	Dual	Egg
Size	Medium	Massive	Medium	Small
Shell colour	Brown	Brown	Brown	White
Examples	1) Rhode Island Red	1) Brahma	1) Cornish	1) Leghorn
	2) Plymouth rock	2) Cochin	2) Australorp	2) Minorca
	3)New Hampshire	3) Langshan	3) Dorking	3) Ancona
	4)Wyandotte		4) Orpington	4) Andalusian
			5) Sussex	

Chicken are also classified based on their utility as

1) Egg type - eg. White Leghorn, Minorca, Ancona
2) Meat type - eg. Cornish, Plymouth rock, Brahma
3) Dual purpose - eg. Rhode Island Red, New Hampshire
4) Game bird - eg. Aseel
5) Fancy variety - eg. Silky, Frizzled, Bantams
6) Desi type - eg. Kadaknath, Naked neck, Chittagong.

Breeds of Poultry

American Class

1. The **New Hampshire** has a more triangular shape than the Rhode Island Red. They have a deep, broad body, feather rapidly, and are prone to go broody, so they make good mothers. They are a dual-purpose breed, laying about 200-220 large, tinted eggs a year. Some strains lay eggs with a dark brown shell color. The American Bantam Association classifies it as a clean legged and single comb breed.

Fig. 3: New Hampshire cock and hen

2. **White Plymouth Rock** Chicken are an esteemed dual-purpose breed. Known for both its high-quality meat and its large brown eggs, this docile breed makes for the perfect backyard chicken. White Plymouth Rocks have a single comb and tolerate both hot and cold climates well.

Fig. 4: White and Barred Plymouth Rock hen and cock

3. The **Rhode Island Red** is an American breed of domestic chicken. It is the state bird of Rhode Island. It was developed there and in Massachusetts in the late nineteenth century, by cross-breeding birds of Oriental origin such as the Malay with brown Leghorn birds from Italy. It was a **dual-purpose breed, raised both for meat and for eggs**; modern strains have been bred for their egg-laying abilities. The colour of the plumage of the traditional Rhode Island red ranges from a lustrous deep pink to almost black; the tail is mostly black. The traditional dual-purpose "old-type" Rhode Island Red lays 200–300 brown eggs per year, and yields rich-flavoured meat.

Fig. 5: Rhode Island Red cock and hen

4. The **Wyandotte** is an American breed of chicken developed in the 1870s. It was named for the indigenous Wyandot people of North America. The Wyandotte is a dual-purpose breed, kept for its brown eggs and its yellow-skinned meat. It is a popular show bird, and has many colour variants. The Wyandotte can be expected to lay approximately 180-200 eggs a year, which makes it a reasonable egg producer. Wyandotte is also an excellent source of meat.

Fig. 6A: Black Laced Silver Wyandotte

Fig. 6 B: Golden Laced Wyandotte

Mediterranean Class

They are light bodied and well developed for high egg production.

1. The **Minorca** is one of the heaviest of the light breeds and originates from the Mediterranean. The breed was developed in England from imported Castilian fowl of Spain. They are utility fowl and were once in the class of widespread large flocks for laying (they lay large eggs) and meat production, like the Leghorn breed which is the smallest of this class. They will not usually go broody. A breed that enjoys foraging

and free-range and will rarely be worried about aerial predators. The distinction of the Minorca is its rather large white ear patch much like the White-Faced Black Spanish (another of this class), which makes it recognizable at a distance. Minorca chickens **lay around 200 to 280 large eggs each year**.

Fig. 7: Minorca cocks and hens

2. Another member of the Minorca group is the **Blue Andalusian** which is noted for its equally Minorca large-sized comb, limiting this breed to warmer regions as frostbite is a problem for these birds. The Blue Andalusian is an incredibly productive breed, laying an average of **250 to 300 eggs per year.**

Fig. 8: Blue Andalusian cock and hens

Classification, Light feather-soft, Appearance, Black, white and blue

Weight: Cock - 3.2 – 4.1 kg, Hen - 2.7 – 3.6 kg

Bantam Variety Minorca Rooster 960 gg, Hen 850 g, Egg shell colour, white

Fig. 9: Bantam Variety Minorca cock and hens

3. The **Ancona** originates from Italy and although this particular bird displays a tendency to be flighty, it is easy to train and makes a great breed for family and backyard producers. After a period of gentle care, Ancona's become a lot more trusting, however, it is still necessary to either clip their wings or ensure that they are completely fenced in. Traditionally, the Ancona variety is not overly broody and if managed correctly can produce up to 250 eggs per year. There is also a bantam Ancona which weighs significantly less than the pure Ancona. The appearance of the Ancona is characterized by colouring that is black (referred to as '**beetle-green**') mottled with white, has spread out toes and yellow legs mottled with black. The Ancona is thought to be related to the Leghorn and in Germany, the breed is called Mottled Leghorns.

   Classification Soft feather – light

   Appearance, the Ancona is beetle green with white tips and a distinctive V-shaped marking on its feathers.

   Large Ancona

   Rooster - 2.7-3.0 kg, Hen -2.25 – 2.5 kg

   Bantam Variety Ancona: Rooster - 570 – 680 g, Hen - 510 – 620 g

   Egg shell colour is white to cream

Fig. 10: Ancona cock and hen

4. **Leghorn** chickens remain perhaps one of the **most popular chicken breeds due to their ability to produce approximately 300 – 320 eggs per year**. Until recently, this breed was the most important in commercial egg production, with **approximately 24 recognized varieties**. Due to their prolific egg-laying ability, they are preferred by laboratories for embryonic and avian biological research as well as being the number one breed **used for large-scale commercial egg production** in the United States. **The Leghorn originates from Italy** and its cross-bred hereditary provides a rarely broody, mobile and efficient scavenging chicken. However, due to their nature, Leghorns do not necessarily make great pets. Leghorns are noisy birds, somewhat smaller than other breeds although they are larger than the bantam.

   Classification, Soft feather – light, Appearance, **there are about 13 colours, ranging from black to blue, brown buff, white to mottled.**

   Weight

   - Cock: 2.8 – 3.4 kg, Hen: 2.0 – 2.5 kg, Cockerel: 2.2 kg, Pullet: 1.8 kg

     Bantam Variety Leghorn

   - Rooster: 1.1 kg, Hen: 0.9 kg, **Egg shell Colour: White**

Fig. 11: White Leghorn cock and hen

Fig. 12: Brown leghorn hen (Single comb)

English Class

1. Cornish: This is the native breed of England. The breed is heavily fleshed (meat) with a compact body and a good depth. Body is well rounded on all sides and carried higher in front than rear (back). The average weight of cock is 4.6 kg and Hen is 3.5 kg. The breed is used as male line in broiler breeding.

Origin: Cornwall, in 1886, a general of the British East India Company claimed that he had developed the breed in Cornwall from Red Aseel he brought from India with Black Red Old English Game. The American Poultry Association (APA) accepted the Dark variety in 1893, and the White in 1898. The APA renamed these "Cornish Indian Game" and "White Indian Game" respectively in 1905. To further align the breed with its origin and qualities, the APA renamed it "Cornish" in 1910, and moved it from the Oriental to the English class.

Description: Broad and deep breast, well-muscled, and compact. Short, thick legs are wide-set. Skull is wide with deep-set eyes, prominent brow, and stout curved beak. Close, short and narrow feathers with little or no down. Tail carried low. Males and females body types are similar, with minor sex differences. Beak and nails are yellow or horn-coloured. Legs are yellow. Wattles and ear lobes are small and red.

Varieties: In the original Dark, the male is mainly glossy beetle-green black with traces of bay; females have black lacing on rich brown. The APA also recognizes **White, White Laced Red, and Buff.** Bantam varieties included are Dark, White, White Laced Red, Buff, Black, Blue Laced Red, Mottled, and Spangled.

In the UK, recognized colours are **Dark, Double-Laced Blue, and Jubilee** (white lacing on chestnut ground). In Europe and Australia, breeders have developed and recognized other colours, such as Blue.

Skin Colour: Yellow

Comb: Pea

Egg Colour: Tinted.

Egg Size: Medium to large.

Productivity: Chicks are slow-growing, ready for harvest in seven months. However, this results in a good quantity of fine, white meat. The hen's muscular body shape limits fertility to about 50–80 eggs per year.

Weight: Large fowl—rooster 10.5 lb. (4.8 kg), hens 8 lb. (3.6 kg); market weight: cockerel 8.5 lb. (3.9 kg), pullet 6.5 lb. (3 kg). UK minimums are 8 lb. (3.6 kg) for males and 6 lb. (2.7 kg) for females.

Bantam—rooster 44 oz. (1.2 kg), hens 36 oz. (1 kg). The Indian Game Club in Britain suggests that bantams do not exceed 4.4 lb. (2 kg) for adult males and 3.3 lb. (1.5 kg) for adult females.

Fig.13: Dark Cornish

Fig. 14: White Laced Red roaster and hen

2. Sussex chicken is a dual-purpose utility breed that is also popular on the show scene and relatively easy to keep. They come in eight colours – **Brown, Buff, Coronation, Light, Red, Speckled, Silver, White (standardized in the UK)**. They are also available in bantam versions.

Sussex chickens are upright, alert, and usually docile. The Sussex Chicken has a wide, flat back, a deep breast and broad shoulders. Their tail is held at 45 degrees. They have a single comb and red ear lobes and comb. They are good foragers, and the Light Sussex is often used to create hybrids. When crossed with a 'gold' cockerel, they will produce sex-linked chicks. They rarely go broody.

The Speckled Sussex was created in Britain in the 18th Century, first appearing at a poultry show in 1845. Other colours were soon created with different crosses, but the Buff Sussex didn't appear until 1920.

Fig. 15: Sussex cock and hen

Utility: Dual purpose, exhibition.

Origin

Eggs: 180 – 210/year, shell colour: cream / light brown.

Weight: Cock: 4.0 – 4.2 kg. Hens: 3.0 – 3.2 kg.

Bantam Cock: 1.0 – 1.2 kg. Hens: 780 – 800 g.

Colours: Brown, Buff, Coronation, Light, Red, Speckled, Silver, White (Standardised UK).

Useful to know: Hardy, good for beginners, rarely go broody. Light Sussex are fast maturing.

3. **The Orphington** fowl is more impressive in the flesh than in photographs that accompany the various books on pure breeds of poultry. With its abundance of feathers, the large fowl Orpingtons fill their show pens and are a sight to behold. The bantams, a miniature version of this magnificent breed, are still relatively big birds and equally eye-catching and impressive.

Viewed by many poultry writers as a 'show breed that does not lay eggs very well,' this label is perhaps a little unfair. After all, the **Guinness Book of World Records for most eggs laid by a pure breed is held by a Black Orpington, who in 1930 laid a phenomenal 361 eggs in a year, so only having four days downtime.**

Fig. 16: Orphington cock and hens

Utility: Exhibition / ornamental.

Origin: Orpington, Kent, Great Britain.

Eggs: 100 – 180 white / tinted.

Weight: Cock: 4.5 kg min. Hens: 3.6 kg min.

Bantam Cock: 2.0 kg, Hens: 1.6 kg.

Colours: **Blue, Black, Buff, White (Standardised United Kingdom) Birchen, Chocolate, Cuckoo, Gold Laced, Jubilee, Lavender, Lemon Cuckoo, Partridge, Red, Spangled** (Non-Standardised).

Useful to Know: Docile, a good choice to have around with children.

4. Australorp is an abbreviation of **Australian black Orpington**, and as the name would suggest, the Australorp is a breed that originated in Australia (in approximately 1890), from the English Orpington. The Australorp is well-suited to Australian conditions and is one of the most efficient egg layers as it averages over **300 eggs per hen per year in a commercial setting**. However, backyard poultry producers should not expect more than 250 eggs a year. The Black Australorp is an ideal bird for free-range production as they have a good temperament. Although they are **renowned for their egg-laying ability, the Australorp is a durable dual-purpose breed as it is also suitable for meat.**

Fig. 17: Black Australop cock and hen

Classification, soft feather-heavy

Appearance, black with a green shine, or pure white

Average weight, Large Australorps

Cock: 3.9 – 4.7 kg, Hens: 3.0 – 3.6 kg, Cockerel: 3.2 – 3.6 kg, Pullets: 3.3 – 4.20 kg

Australorp, female, Bantam Australorps

Rooster: 1.2 kg, Hens: 790 g, Cockerel: 1.6 – 2.1 kg, Pullets: 1.3 – 1. 9 kg

Egg shell colour, light brown/tinted

Asiatic Class

The common characteristics of birds belonging to Asiatic class of chicken are presented in Table 7.

Table 7: Common characteristics of Asiatic class of chicken

<table>
<tr><th rowspan="3">Characteristics</th><th colspan="4">Colour of</th><th rowspan="2">Comb</th><th rowspan="2">Shank feathering</th></tr>
<tr><td>Ear lobes</td><td>Skin</td><td>Shank</td><td>Eggs</td></tr>
<tr><td>Red</td><td>Yellow[1]</td><td>Yellow[2]</td><td>Brown</td><td>Single[3]</td><td>Present[4]</td></tr>
<tr><td>Breeds</td><td colspan="6">Varieties</td></tr>
<tr><td>Brahma</td><td colspan="6">Light, Dark, Buff</td></tr>
<tr><td>Cochin</td><td colspan="6">Buff, Partridge, White, Black, Silver-laces, Golden-laced, Blue, Brown, Barred</td></tr>
<tr><td>Langshan</td><td colspan="6">Black, White, Blue</td></tr>
<tr><td colspan="7">[1]White in Langshan
[2]Bluish-black in Langshan
[3]Pea comb in Brahma
[4]Characteristic feature of the Class</td></tr>
</table>

Some of the important breeds of this class are as follows:

1. **Brahma:** It is originated from **Brahmaputra region** of India, where they are known as "Gray Chittagongs".

   - The Brahma chickens are a cross breed from the large feather legged birds known as Shanghais from China in the 1840s by Grey Chittagongs from India and produced the pea comb Brahma that we see today.

   - The body is circular shaped and massive in appearance due to profuse, loose feathering and feathered legs and toes. The mature weight of cock is 5.5 kg and Hen is 4.2 kg.

- Modern-Day Brahma Breed: Today, the Brahma breed is still loved and admired for its unique characteristics. It comes in three recognized varieties: **light, dark, and buff.**

Fig. 18: Brahma hens and cock

1. **Cochin:** It is **originated from Cochin in India**. It is a fancy bird. The standard weight of cock is 5.1 kg and hens is 3.9 kg.

Fig. 19: Cochin cock and hen

Origin

- Cochin chickens were initially known as "Shanghai" or "Cochin-China" chickens.
- They were named after the **port city of Cochin in India**, from where they were often shipped.
- These chickens originated in China and were brought to Europe and the United States in the mid-19th century.
- Their unique appearance and large size quickly gained popularity among chicken enthusiasts.

Physical Characteristics

- Cochin chickens have a huge body with plenty of feathers. They are known for their feathered legs.
- The breed comes in various colors, including black, blue, buff, cuckoo, partridge, grouse, and white.
- Cochin males weigh between 3.6–5.9 kg (standard) and 900 g (bantam), while females weigh between 3.2–5.0 kg (standard) and 800 g (bantam).
- They lay brown eggs and produce around 150–180 eggs annually.
- Cochin chickens have a single comb and are hardy in winter.

Behavior and Personality

- Cochin chickens are calm, docile, and friendly birds, making them ideal for pet lovers.
- While they are not prolific egg layers, they excel as good mothers.
- Their friendly personalities endear them to their owners.
- Cochin chickens are disease-resistant and hardy in winter climates.

2. **Langshan:** Langshans **originated in China from a district around the Yangtze Kiang River near Shanghai.** The Langshan is considered one of our oldest breeds. The breed became popular at the height of the "hen fever" with the Cochins and Brahmas.

Fig. 20: Langshan cock and hen

- The breed gets its name from the Langshan Mountains, where they were first discovered.
- Langshans are large birds, with roosters typically weighing in at around 9.5 pounds and hens around 7 pounds. They are known for their long, elegant necks and tails, as well as their black plumage.
- All varieties have **white skin, but black, white and blue feathered varieties have emerged over the years**. The "official" varieties of Langshan in the U.S. are black, white, and blue.

Table 8. Indian breeds of chicken: Currently 20 chickens are register by ICAR-NBAGR.

Sl. No.	Breed	Home Tract
1.	Nicobari	Andaman & Nicobar
2.	Danki	Andhra Pradesh
3.	Kalasthi	These are native to the Chittoor district, and adjoining regions of Nellore district in Andhra Pradesh.
4.	Ghagus	Andhra Pradesh and Karnataka
5.	Daothigir	Assam
6.	Miri	Assam
7.	Aseel	Chhattisgarh, Orissa and Andhra Pradesh
8.	Ankleshwar	Gujarat
9.	Busra	Gujarat and Maharashtra
10.	Kashmir Favorolla	Jammu and Kashmir
11.	Tellicherry	Kerala
12.	Kadaknath	Madhya Pradesh
13.	Kaunayen	Manipur
14.	Chittagong	Meghalaya and Tripura
15.	Hansli	Odisha
16.	Punjab Brown	Punjab and Haryana
17.	Mewari	Rajasthan
18.	Uttara	Uttarakhand
19.	Haringhata Black	West Bengal
20.	Aravali	Gujrat

Description of indigenous breeds

1. Aseel

- **Synonyms: Peela, Noorie, Chitta, Yakub, Kagar, Java, Sabja, Teekar and Reja.**
- **The Asil or Aseel** is an Indian breed or group of breeds of game chicken used for cock fighting and meat purposes. It is distributed largely in India, particularly in the states of **Tamil Nadu, Andhra Pradesh,**

Chhattisgarh and Odisha. It is also present in Bangladesh and Pakistan, which were part of India until Partition. This breed has been exported to several other countries like Australia, Guatemala, Honduras, Ireland, Luxembourg, the United Kingdom, the United States and Uruguay. Similar fowls are found throughout much of Southeast Asia.

- The name "Aseel" (or "Asil") means "pure" in Arabic and "original," "high caste," "high born," or "pure" in Hindi.
- Originating from the Indian subcontinent (including Sri Lanka, Pakistan, Bangladesh, and present-day India), Aseels have an ancient lineage.
- It is one of the parent breeds of the **Indian Game**, developed in the West Country of England and USA in the early nineteenth century.

Fig. 21: Aseel breed of hen and cocks

Characteristics: There are many varieties of Asil. Among them are the Amroha, Bhaingam, Kilimooku, Kulang, Lasani, Madras, Mianwali, Reza and Sindhi types.

Aseel (Asil) birds are multi-coloured plumage birds, their primary colours being dark brown, golden, black, and other colours

- Some of the most popular Aseel chickens include White (Heera), Golden-red (Peela), Light red (Reza), Brown (Teekar), Black and white-silver (Chitta), Black and red (Yarkin and Amroha), Black (Kagar), Black with green shades (Mushka and Shamo)
- Aseel typically live for approximately 10 years.
- These birds have distinctive appearances are long, slender face (not covered with feathers), compact eyes, long neck, small tail, strong, straight legs.

The standard weight for Aseels is as follows

Cocks: 4-5 kg, Hens: 3-4 kg, Cockerels: 3.5-4.5 kg, Pullets: 2.5-3.5 kg

Use: Asil hens are not good layers, but sit well. They may lay about 70 eggs per year; the eggs vary from cream-coloured to brownish, and weigh approximately 40 gm.

The population of Aseel breed in our country is 3,36,80,583 as per 20th Livestock Census, 2019 which is the highest 4.2 % share of breed with respect to total in India.

2. Kadaknath, also called Kali Masi ("fowl having black flesh")

Kadaknath, also called Kali Masi ("fowl having black flesh"), is an Indian breed of chicken. They originated from Dhar and Jhabua, Madhya Pradesh. These birds are mostly bred by the rural and tribals. There are three varieties: jet black, golden and pencilled. The meat from this breed has a geographical indication (GI Tag) tag that was approved by the Indian government on 30 July 2018.

Fig. 22: Kadaknath chicks

- The Kadaknath is popular for its adaptability and its grey-black meat, which is believed to infuse vigour. Its colour is caused by melanin. The breed is considered to have originated from the Kathiawar, Alirajapur jungles in Jhabua district of Madhya Pradesh.
- The roosters weigh 1.8–2 kg (4.0–4.4 lb) and the hens 1.2–1.5 kg (2.6–3.3 lb). Kadaknath hen's eggs are brown with a slightly pink tint; they are poor setters and rarely hatch their own brood. The egg shell colour most frequently is dark brown (67.87%) followed by light brown colour around 32.12%. Eggs weigh an average of 40 to 49 g. The yolk colour is medium dark yellow or orange. Average annual egg production of Kadaknath chicken is around 80 to 120 eggs.

Colour: Kadaknath birds are grey-black all over and have gold plumage with greenish iridescence. The greyish black colour is present in the legs and toenails, beak, tongue, comb and wattles; even the meat, bones, blood and organs have black coloration.

Fig. 23: Kadaknath hen and cock

Morphological characters

Adult weight (avg. kg): Male 1.6 kg, Female 1.125 kg

Plumage colour: Ranges from silver to gold spangled to blue black

Skin colour: Dark grey, Shank colour: Grey

Egg shell colour: Light Brown

Visible character: The colour of day-old chicks is bluish to black with irregular dark stripes over the back. In the adults, comb, wattles and tongue are purple. The shining blue tinge of the ear lobes adds to its unique features.

Performance: Annual egg production (avg. no.): 80.0, Egg weight (avg. gm): 40.0

Threat of extinction

Due to the relatively high consumption of the breed, its numbers have sharply declined throughout the years. To save the breed from extinction, the state government started a Kadaknath poultry breeding program involving 500 families who are below the poverty line, who were to receive financial support and assistance.

The population of Kadaknath breed in our country is 20,37,475 as per 20th Livestock Census, 2019.

- Most of the internal organs show the characteristic black pigmentation which is more pronounced in trachea, thoracic and abdominal air sacs, gonads, elastic arteries, at the base of the heart and mesentery.
- The black colour is due to the deposition of Melanin pigment. (Fibromelanosis). The meat and eggs are reckoned to be a rich source of protein and iron.

- Resistant to diseases in its natural habitat in free range but is more susceptible to Marek disease under intensive rearing conditions.
- Comments on utility: The black flesh is considered not only a delicacy but also of medicinal value. The tribals use Kadaknath blood in the treatment of chronic diseases in human beings and its meat as aphrodisiac. It is considered to be a sacred bird and offered to the Goddess after Diwali.

3. Nicobari fowl is well adapted in the island milieu condition under impending climate change scenarios. Nicobari fowl in Mon-Khmer language (Austro-Asiatic language spoken by Nicobarese) is called Takniet, which means 'short legged chicken'. Nicobari fowl is an endemic breed of Nicobar group of Islands and maintained their unique genetic identity due to natural geographical barriers.

- The one and unique breed has evolved in the Nicobar group of islands and is genetically closely related to Red jungle fowl and belongs to *Gallus domesticus* species. They are medium sized with short legs. They have compact body conformity. These birds are mostly single combed.
- Wattles and ear lobe are pinkish in colour. They have short and thick necks with mostly black plumage tipped with brown shade, breast bulging in front, medium size tail and long saddle feathers fitting well with the tail.
- There are three strains of Nicobari fowl, namely, Brown, Black, and White.
- **Nicobari fowl is less susceptible to Ranikhet disease**, Mareks disease and Infectious bursal disease, Salmonella, *E. Coli*, and Coccidiosis unlike exotic breeds of fowl.
- They are comparatively better in egg productivity compared to desi birds reared in backyard farming. They produce 130-140 eggs per annum and gain 1.2 kg body weight at 20 weeks of age.
- The population of Nicobari fowl breed in our country is 26,615 as per 20th Livestock Census (2019).

Fig. 24: Nicobari cock and hen

4. Danki

Home Tract: Vizianagaram, Visakhapatnam and Srikakulam districts of Andhra Pradesh. Local people call these birds by different names on the basis of plumage colour: - **Black coloured birds** - khaki or sanwla, red colour-dega, brick colour-parla, white-satua and spotted-pingle.

Morphological characters

It is mainly fighting or game birds where Danki cocks are used for Danki fights using their natural heels without slashers during Sankranti. This breed is very similar to Aseel in morphological characteristics, but Danki has a shorter and stronger neck, broader wings, and sharp eyes.

Body: Fairly large, medium sized, round

Body Plumage (Frizzled/Normal)

Colour: Highly variable. In most of the birds' plumage colour is brown all over the body with a few black feathers on the ventral surface. In some birds it is glossy, lustrous and multi-coloured varying from intensive black with white mix to complete black. Cocks generally have shining bluish black feathers on wings, breast, tail and thighs.

Pattern: Patchy in males and spotted in females

Skin colour: Varies from pink to light red

Head: Fairly large, broad, deep

Combs Size: Large, compressed and is positioned high on the head

Type: Pea

Colour: Brick red

Wattles: Absent or small, brick red in colour

Ear lobe colour: Oval in shape, medium sized, red in colour

Morphometric traits:

Body weight: Cock 3.115±0.0922 kg, hens 2.23±0.064 kg

Shank length: Cock 11 cm, hesn 9 cm

Reproductive Traits: Age at first egg: 6-8 months

Hatchability on total egg production: 70-85%, Broodiness: Usual

Production traits: Egg shell colour: Light Brown, brown, dark brown

Annual egg production (numbers): 25-35, Weight of an egg: 37-54 g

Climatic adaptation: Well adapted to harsh environmental conditions, more resistant to internal parasitic infestations.

The population of Danki breed in our country is 1,68,332 as per 20th Livestock Census (2019).

Fig. 25: Danki hen and cock

5. Kalasthi

Home Tract: **Chittoor, Cuddapah and Nellore districts of Andhra Pradesh**. This breed might have been named after the name of the area i.e. SriKalahasti in Chittoor district where these birds are found. However, in the breeding tract, the birds are known not by the name Kalasthi but as desi.

Hardiness: Kalasthi chickens are well-adapted to their environment and have been intricately associated with the culture of the farmers and breeders who maintain them.

Entertainment and Betting: Kalasthi chickens are also used in cockfights, a common practice among the tribals in the region.

Morphological characters

Body: These birds **resemble Danki birds except that they are smaller in size**

Body Plumage: Bluish black or brown; Peacock type bluish in colour

Plumage pattern: Patchy

Skin colour: White or Pinkish

Type comb: Pea or Single, **Comb colour:** Red

Ear lobe colour: Red, Eye ring colour: Red

Fig. 26: Kalasthi cock and hen

Neck: Long and is covered with golden feathers. Brown coloured birds have dark brown feathers on their necks. Cocks have shining bluish black feathers.

Wings: Dorsally set exposing thighs which are covered with smooth feathers

Tail: Brown coloured birds have bluish black or dark brown on tail

Limbs/Shanks

Colour: Grey or Yellow, **Spur:** Small

Morphometric traits: Body weight: Cock 2.482±0.13 kg, hens 1.85±0.102 kg

Reproductive Traits: Age at first egg: 5-9 months. Hatchability on total egg production: 60-85%

Broodiness: Usual

Production traits: Egg shell colour: Light Brown, brown, Dark Brown

Annual egg production (numbers): 30-42, Weight of an egg: 42.91±1.94 g

Utility: These birds are mainly kept for meat purposes and cocks are occasionally used for fighting. The consumption of eggs of desi birds is very little in this area owing to high returns from the sale of chicks. The population of Kalasthi breed in our country is 49,736 as per 20th Livestock Census (2019).

6. Ghagus

Class: Asiatic, Country: India

Home Tract: Kolar and Bangalore districts of Karnataka, Chittoor and Anantapur districts of Andhra Pradesh. This breed might have derived its name from its peculiar sound 'Ghegu' and the birds are locally known as Desi. These birds look **mixture of Aseel type and Desi egg laying birds**.

They are **smaller in size compared to breeds like Aseel, Danki, and Kalasthi**, but their egg production surpasses that of these other breeds.

Morphological characters

Body: Smaller in size as compared to Aseel, Danki and Kalasthi

Body Plumage: Brown or black; Cocks have shining bluish black feathers on breast, tail and thighs. Neck is covered with golden feathers.

Plumage pattern: Patchy

Skin colour: White

Combs Size: Medium Type: Pea or Single Colour: Red

Wattles: Small and red in colour.

Ear lobe colour and Eye ring colour: Red

Neck: Throat in some cases is loose and hanging.

Limbs/Shanks: Yellow in colour

Morphometric traits: Body weight: Cock 2.16±0.25 kg, hens 1.433±0.81 kg

Fig. 27: Ghagus cock and hen

Reproductive Traits: Age at first egg: 5-8 months. **Hatchability on total egg production:** 60-85%, **Broodiness:** Usual

Production traits:

Egg shell colour: Light brown, brown, **Annual egg production (number):** 45-60

Weight of an egg: 40.25±2.39 g

Rearing system: Housing of birds is open. Sometimes the shelter is provided under the dry fodder stack which is kept at a height of about 1.0 to 1.5 feet above the ground on stone pillars. Birds use this space to keep away from sun and predators. Cocks are tied by a rope and steel strips are used on both ends of rope to prevent coiling.

Utility: Mainly kept for meat and egg purpose.

The population of Ghagus fowl in our country is 7,60,936 as per 20th Livestock Census (2019).

7. Daothigir

Class: Asiatic, Country: India

Daothigir chickens are primarily found in the **Bodoland region of Assam**, specifically in districts such as Kokrajhar, Chirang, Udalguri, and Baska.

Home Tract: Kokrajhar, Bongaigaon, Barpeta, Dhubri and Nalbari districts of Assam. The name "Daothigir" is derived from the Bodo language, where "Dao" means bird. The breed derives its name from the name of a plant -Thigir

(*Dillenia indica*) found in this region. This plant bears flowers of **different colours similar to the plumage colour of these birds**. The shape of these flowers also resembles the comb of these birds. In Bodo language Dao means bird and hence these birds are known as Daothigir.

Morphological characters

Body: Small sized, compact but heavy

Body Plumage: Frizzled/Normal

Colour: Black interspersed with white feathers, **Pattern:** Striped or Spotted

Fig. 28: Daothigir cock and hen

Skin colour: Creamish slightly towards pinkish

Combs, Size: Large, erect. **Type:** Single Colour: Red

Wattles: Medium to large in size, red coloured

Ear lobe colour: Red, **Eye ring colour:** Red

Neck: Golden yellow or brown feathers in brown coloured birds

Wings: Black or brown feathers

Back: Golden yellow or brown feathers in brown coloured birds

Tail: Black or brown feathers; short and almost in level with the back

Limbs/Shanks: Long, **Colour:** Yellow

The average flock size in the Daothigir breeding area ranges from 10 to 60 birds.

Male birds are often castrated (caponized) at around 2 to 4 months of age to enhance growth and meat quality.

Morphometric traits

Body weight: Cock 1.792±0.129 kg, hens 1.625±0.126 kg

Reproductive traits: Age at first egg: 5-8 months.

Hatchability on total egg production: 80-85%

Broodiness: Usual

Production traits

Egg shell colour: Light Brown, Brown, Dark Brown, Creamy

Annual egg production (number): 60-70, **Weight of an egg:** 44.42±1.35 g

Remarks: Male birds are castrated (caponized) at about 2–4 months of age for fattening. This is done with a belief that caponized birds have faster growth and better meat quality as compared to the entire ones.

Utility: These birds are reared mainly for meeting the domestic requirements of meat and eggs. These also serve as a cash reserve for the farmers.

The population of Daothigir fowl in our country is 36,96,257 as per 20th Livestock Census (2019).

8. Miri

Class: Asiatic, Country: India

- **Home Tract: Sibsagar, Lakhimpur, Dhimaji, Dibrugarh and Majhauli districts of Assam**. The name **Miri is derived from the name of the tribe 'Miri'** (now missing tribe) who rear these birds.
- The Miri chicken, also known as **Porog**, is a native breed found in the northeastern state of Assam.
- On average, each household maintains a flock of 25.2 birds, consisting of 11 males and 14 females.
- Miri chickens make up 90-95% of the flock.
- These birds are reared as backyard farming by female family members.
- They are a scavenging type of chicken and require minimal input for survival and reproduction.

Morphological characters

Body Plumage: Frizzle in very few birds

Colour: White or brown or black

Plumage pattern: Mostly solid, sometimes dull or patchy or spotted or striped

Skin colour: White, Combs Size: Large in cocks and small in hens

Comb type: Single**, Colour:** Red, Ear lobe colour: Red

Eye ring colour: Reddish brown

Neck: Naked in very few birds

Limbs/Shanks: White in colour

Morphometric traits:

Body weight: Adult 1.525 ± 4.95 kg

Reproductive Traits: Age at first egg: 6.58-7.57 months, **Fertility rate:** 87-91%

Hatchability on total egg production: 73-83%, **Broodiness:** Usual

Production traits:

Egg shell colour: Light brown to brown, **Annual egg production (numbers):** 54-67

Weight of an egg: 42.06±0.17 grams, **Dressing percentage:** 65-74%

Food conversion ratio: 2.91

Utility: Meat and eggs. The tribals use these birds invariably in their social and religious rituals.

Fig. 29: Miri cock and hen

The population of Miri fowl in our country is 54,11,093 as per 20th Livestock Census (2019).

9. Ankleshwar

Home Tract: Bharuch and Narmada districts of Gujarat. The named with the place where it has been bred i.e. **Ankleshwar** in Bharuch district of Gujarat. The breeding tract ranges to Jumbusar, Zagadia, Bharuch, Hansot and Valia of Bharuch and Dediapada, Rajpipla, Tilakwada and Nadod of the Narmada districts of Gujarat.

Morphological characters:

Body: Small to medium sized bird; Body Plumage: Frizzled/Normal

Colour: Golden yellow or yellow with black stripesor white and golden with black stripes. Golden yellow plumage is predominant in cocks while Black golden is more common in hens. Feathered legs, cap feathers and bearded feathers are also observed in some of the birds.

Pattern: Striped/speckled or Spotted or Solid

Skin colour: Yellow to pinkish

Fig. 30: Ankleshwar cock and hen

Combs Size: Large in cocks and small in hens

Type: Single or Rose, **Colour:** Red

Wattles: Red, **Ear lobe colour**: White

Eye ring colour: Yellow reddish

Limbs/Shanks: Yellow or black in colour

Morphometric traits

Body weight: Cock 1.759±0.007 kg, hens 1.487±0.006 kg

Reproductive Traits

Age at first egg: 5.92±0.008 months, **Fertility rate:** 91.3%.

Hatchability on total egg production: 92.4%. **Broodiness:** Usual

Production traits

Egg shell colour: Cream/Brown/White

Annual egg production (numbers): Poor egg production (79.35±0.291 per annum).

Weight of an egg: 34.3 g, **Age at slaughter:** 16.59±0.073 months

Slaughtering weight: 1.687 kg, **Dressing percentage:** 62.44%

Remarks: This breed is mainly **reared by tribal communities in South Gujarat** for backyard poultry farming. They are maintained without vaccination and medication, and have reasonable feed efficiency as they survive on 25-30g of grains, scavenging and maintain excellent fertility.

Utility: Meat and eggs.

The population of Ankleshwar fowl in our country is 13,22,006 as per 20th Livestock Census (2019).

10. Busra

- **Home Tract: Dhule, Nandurbar, Nasik districts of Maharashtra, Surat, and the Dangs in Gujarat.** Local birds in the Busra breeding tract are not known by the name Busra but as Desi. Busra nomenclature might have come from **Busrawal, a village in Sakri taluka of Dhule district or from the name of a tree 'Busrawal' (Marathi) / Bahawa (Advasi dialect).**
- Locally, these birds are known as "Desi" rather than Busra.

- Interestingly, this breed is exclusive to backyard poultry in Maharashtra and plays a significant role in the tribal economy.

Morphological characters

Body: Small in size

Body Plumage: White mixed with black. There is a wide variation in plumage colour which is mostly white mixed with black feathers on neck, back, tail, and reddish-brown feathers on shoulders and wings. Good number of birds have white plumage mixed with light brown feathers throughout. Some are solid white in colour. Birds having brown mixed with black or black mixed with golden feathers or solid black plumage are also available.

Pattern: Spotted, **Skin colour:** Pinkish

Combs size: Small to medium in size, stands erect, **Type:** Single type, **Colour:** Red

Fig. 31: Busra cock and hen

Wattles: Red, **Ear lobe colour:** White/brown

Eye ring colour: Red, **Beak:** Yellow in colour

Limbs/Shanks: Yellow in colour

Morphometric traits

Body weight: Cock 0.85-1.25 kg, hens 0.8-1.2 kg

Reproductive Traits

Age at first egg: 5-7 months, **Hatchability on total egg production:** 60-85%

Broodiness: Usual

Production traits

Annual egg production (annual): 40-55, **Weight of an egg:** 28-38 g

Dressing percentage: 65-70%

- **Busra chickens are kept in a free-range system.**
- Flock sizes vary, with an average of 8.6 birds per flock.
- The composition of a typical flock includes 48% chicks, 39% hens, and 13% cocks.
- Shelter is primarily provided to chicks in the form of bamboo baskets

Utility: Meat and eggs. These birds are reared for home consumption as well as for sale of live birds and eggs.

- The population of Busra fowl in our country is 12,18,724 as per 20th Livestock Census (2019).

11. Kashmir Favorolla

Home Tract: Srinagar, Baramulla, Anantnag, Budgam, Kupwara, Pulwama districts of Jammu and Kashmir. Kashmir Favorolla is a native chicken breed primarily reared in the high-altitude regions of the Kashmir valley in Jammu and Kashmir, India.

Morphological characters

Body Plumage: Silky frizzled

Colour: Mixed shades of Black, Red, Green, Gold

Plumage pattern: Solid, stripped, Patchy, Spotted, Barred

Skin colour: White

Crested/Plain: Feathered cap - tuft of feathers on head.

Combs Size: Large in cocks and small in hens

Type: Single, **Colour:** Red

Ear lobe colour: White, **Ear tufts:** Present

Eye ring colour: Red,

Neck: Naked, **Limbs/Shanks:** Feathered, **Colour:** Yellow, **Spur:** Multiple

Fig. 32: Kashmir Favorolla cock and hen

Morphometric traits

Body weight: Cock 1.875±0.318 kg, hens 1.415±0.311 kg

Reproductive Traits

Age at first egg: 6.9 months, **Fertility rate:** 77%, **Hatchability on total egg production:** 64%, **Broodiness:** Usual

Production traits

Egg shell colour: Light to dark brown, **Annual egg production (number):** 60-85, **Weight of an egg:** 45.76±2.188 grams, **Remarks:** This breed is most suitable for cold climates and mountainous terrains.

Notable features include

Feathering: Most birds are normally feathered, but some may have a naked neck or a bottle jaw.

Head cap: Many birds sport a feathered cap on their heads.

Comb types: Common comb types include single, pea, rose, and walnut.

Peculiarity of the breed: The typical features are **feathered cap-tuft of feathers on head**. The other physical traits at varied frequencies are: feathered shanks, **tuft of feathers over earlobes**, naked neck, silky frizzle, multiple spurs.

These birds are known for their **hardiness** and ability to survive and produce during subzero temperatures.

They exhibit resistance to various bacterial and infectious diseases.

Locally, they are referred to as "Kashir Kukkar"

Utility: The birds are reared for meat and egg production and constitute one of the major sources of animal protein. It is more important because the commercial poultry farming could not make much headway due to extremes of temperature and increased maintenance cost.

- The population of Kashmir Favorolla fowl in our country is 5,97,626 as per 20th Livestock Census (2019).

12. Tellicherry

- **Home Tract: Kozhikode (Calicut), Kannur Malappuram districts of Kerala and Mahe (Pondicherry).**
- The breed is named after the place Tellicherry (also known as Thalassery) in Kannur district of Kerala.
- **Tellicherry chickens are primarily found in Kozhikode district**, with a few also present in the surrounding areas of Kannur and Malappuram districts.
- These birds are fast-moving and less susceptible to predators.
- They are **mainly reared for meat consumption**.
- Interestingly, Tellicherry birds are believed to have medicinal value. Their meat is used in the preparation of ayurvedic medicines for treating asthma, anemia, and worm infestations.

Morphological characters

Body Plumage: Black with shining bluish tinge. Plumage colour is black with shining bluish tinge on hackle, back and tail feathers. Few birds have golden mixed with bluish feathers on the neck.

Plumage pattern: Solid, **Skin colour:** Grey

Combs Size: Large, erect in cocks and drooping on the rear side in hens, **Type:** Single, **Colour:** Red.

Ear lobe colour: Red, **Eye ring colour:** Blackish red

Limbs/Shanks: Blackish grey in colour

Morphometric traits

Body weight: Cock 1.62±0.16 kg, hens 1.24±0.10 kg

Fig. 33: Tellicherry cock and hen

Reproductive Traits

Age at first egg: 5-8 months

Hatchability on total egg production: 70-80%

Broodiness: Usual

Production traits

Egg shell colour: Light brown, brown, creamy

Annual egg production (number): 60-80

Weight of an egg: 40.02±0.94 g

Remarks: Hen makes a lot of sound after laying eggs. The eggs are tinted and small to medium in size.

Utility: These are reared mainly for meat. These birds are also thought to have some medicinal value. These are used for preparation of Ayurvedic medicines for asthma treatment. Its soup is also believed to be beneficial for treatment of anaemia and worm infestation.

- The population of Tellicherry fowl in our country is 1,48,705 as per 20th Livestock Census (2019).

13. Kaunayen

- **Home tract:** Distributed in the whole of the **Imphal valley** comprising Thoubal, east and west Imphal and Bishnupur districts of Manipur. Some birds are also available in hill regions consisting of Chandel, Churachandpur, Senapati, Ukhrul, Tamenglong districts.

- These chickens are primarily reared in the backyard system, mainly for **cockfighting**.
- **Origin and name:** The name "Kaunayen" is derived from two Manipuri words: "Kauna," which means "kick" or "fighting," and "yen," which translates to "hen" or "poultry."

Morphological characters

Body: Elongated

Body plumage: Predominant plumage colour is black followed by brown or red. Cocks generally have shining bluish feathers on wings, breast, tail and thighs. Hens are generally black, grey, blackish grey or whitish grey with few brown feathers on neck, breast and wings.

Plumage pattern: Generally patchy in males and solid in females

Combs: Pea comb, red in colour

Ear lobe colour: Red, **Eye ring colour:** Red

Neck: Long, generally bare, hard and rose coloured in fighting cocks

Chest/Breast: Generally bare, hard and rose coloured in fighting cocks

Thigh: Generally bare, hard and rose coloured in fighting cocks

Limbs/Shanks: Long, Spur: Long and sharp in cocks

Fig. 34: Kaunayen cock and hen

Morphometric traits

Body weight: Cock - 2.4-3.8 kg, hens - 1.0-2.9 kg

Reproductive Traits

Age at first egg: 5-7 months. **Hatchability on total egg production:** 65-100%

Broodiness: Usual

Production Traits

Egg shell colour: Brown, **Annual egg production (number):** 35

Weight of an egg: 42.43±0.7 g

Temperament: Fighting and can fight for longer duration.

Climatic adaptation: Better adapted to local climate

Rearing system: Birds are reared in backyard system and managed mainly by men. Some breeders rear these birds especially cocks under intensive system. Scavenging with supplementation of kitchen waste and local feeds is the most common feeding system and no commercial feed is given.

Remarks: Morphologically, looks similar to Danki and Aseel birds.

Utility: These birds are mainly reared for cock fighting and contribute a lot in generating income for the poultry keepers.

- The population of Kaunayen fowl in our country is 1,14,828 as per 20th Livestock Census (2019).

14. Chittagong

Synonym: Chittong, Malay

Home Tract: North Eastern states of India, Meghalaya and Tripura, bordering Bangladesh.

- A large bird very strong and hardy with a quarrelsome temperament. Possesses all the characteristics of a **good game bird** such as colouring (primarily bay, chestnut, gray, roan, palomino, black, etc.), large eyes, long mane and tail, strong, yet refined legs, high headset when in action, and low tail set.
- Adult Chittagong birds are strong, hardy, and quarrelsome. They serve as good game birds and are **suitable for both meat and egg production**.

Morphological characters:

Body: Large

Shoulders: Broad with slight narrow loins.

Body plumage: Frizzled/Normal

Plumage (colour): Standard plumage colour is lacking, but the buff, white, black, dark brown and grey varieties are recognized.

Plumage pattern: The plumage is close to the body, firm, short and glossy.

Head: Long, high headset when in action.

Comb: Small, pea comb resembling a small lump of tiny warts, red in colour.

Wattles: Red and hardly visible in the hens.

Ears: Small, usually red and at times admixture with a little white. They are prominent and over-hanging.

Eyes: Eye ring colour, **Beak:** Long and yellow, **Neck:** With long mane

Wings: Projected at the shoulders and are carried high, **Tail:** Long with low tail set

Chest/Breast: The breast is broad, deep and fleshy, **Limbs/Shanks:** Yellow and featherless

Fig. 35: Chittagong hen and cock

Morphometric Traits

Body weight: Cock 3.5-4.5 kg, Hen 3-4 kg

Height: Cocks 75 cm from beak to toe.

Reproductive Traits

Broodiness: Poor mothering ability

Remember, these chickens are not commonly seen in commercial poultry due to their slow maturity, but they **lay approximately 140 eggs per year**.

Remarks: The adult birds are very strong, hardy and quarrelsome. **Good game bird and it is a dual-purpose breed.**

- The population of **Chittagong** fowl in our country is 12,41,000 as per 20th Livestock Census (2019).

15. Hansli

Synonyms: Kalua, Dhobla, Rangua, Nalia, Khadia, Jhinjiria.

Home Tract: Keonjhar and Mayurbhanj district in Orissa.

The birds are distributed throughout the Mayurbhanj district but predominantly found in Khunta, Kaptipada, Udala, Suliapada, Shyamakhunta, Thakurmunda, Bangiriposhi, Saraskana and Kuliona blocks and adjacent areas of Patna and Sarpada blocks of Keonjhar district. Breeding tract is mostly confined to areas surrounding the Similipal national park and biosphere reserve.

Origin

The name Hansli seems to have originated from the word "hans" which means swan. Due to the tall size and majestic gait of the birds like that of swan, the local people call these birds Hansli. Birds are priced very high due to their fighting ability.

Morphological characters

Body: Fairly long, rectangular, deep and well rounded.

Plumage: Predominantly black in colour. The plumage colour of body and breast is dark grey. Under colour of the body including breast and abdomen are slaty or dark steel grey. Hackle feathers are rich golden yellow or red hackle and saddle feathers in cocks and light golden yellow in hens. Primary feathers are light yellow, secondary feathers are dark grey, and coverts of the primaries are black. In some of the hens, primary and secondary feathers are red in colour. Tail feathers are lustrous black with greenish sheen. In some of the bird's plumage colour on the ventral side of the body is black which extends up to hock joint. Plenty of hackle feathers flow over the shoulder in males.

Wattles: Small and rudimentary and red in colour

Ear lobe colour: Medium to large sized, oval, smooth, red in colour

Morphometric traits

Body weight: Cock 2.5-3.8 kg, hens 1.75-2.5 kg, cockerels 1.5-2.0 kg, pullets 1.2-1.4 kg

Height: Cock 61-75 cm, hens 45-63 cm, **Body length:** Cock 71-74 cm, hen 48-57 cm

Shank length: Cock 11.4-13.3 cm, hens 7.2-8.4 cm, **Shank width:** Cock 10.7-15.4 mm, hens 8.9-12.3mm, **Average width of head:** Cock 34 mm, hens 29 mm, **Keel bone length:** Cock 19.0-23.0 cm, hens 10.5-16.1cm

Fig. 36: Hansli breed of cock and hen

Reproductive Traits: Age at first egg: 6 months.

Production traits

Egg shell colour: Light brown, **Annual egg production (number):** 50-67

Weight of an egg: 40-46 g but in some cases 52-58 g

Utility: All the eggs produced are hatched and rarely used for consumption, unless the weather conditions are adverse for a good hatch. Meat fetches a good price in the market since it is considered tastier than that of farm-bred chicken. Fighting cocks are sold at exorbitant prices, usually 3 to 4 times higher than when they are sold for meat purposes.

The population of **Hansli** fowl in our country is 91,385 as per 20th Livestock Census (2019).

16. Punjab Brown

Home Tract: Punjab Brown breed birds are found in rural areas of Punjab and Haryana. They are used for both meat and egg production. The breed seems to have taken its name from its native tract (Punjab) and its colour (Brown).

Some black or white coloured birds with a golden colour on their neck, wings and tail are also available. The pattern is usually solid but sometimes it is spotted or striped. Males in particular have black spots/stripes on their neck, wings and tail. The neck is darker in colour (brown/golden) than the rest of the body.

The birds are reared in a backyard system. Shelter is provided mostly during night. About 10% of farmers keep the birds confined both during the day and at night. Enclosures are small, mostly made of mud (68%). About 30% were made of bricks and 2% of wood. Most of the enclosures were single storied and only about 2% were multi-storeyed. Some of the farmers have even made provision for birds below the mangers of cattle or buffalo. Chicks are kept under a basket made of bamboo sticks.

Morphological Characters

Body Plumage: Brown. Males usually have black spots/stripes on neck, wings and tail.

Plumage pattern: Solid, sometimes spotted or striped. **Skin colour:** White

Combs Size: Large, **Type:** Single, **Colour:** Red, **Wattles:** Red, large sized in males and small in females, **Ear lobe colour:** Brown, **Eye ring colour:** Red, **Neck:** Darker in colour (brown/golden) than the rest of the body, **Limbs/ Shanks:** Yellow in colour

Fig. 37: Punjab Brown hen and cock

Morphometric Traits

Body weight: Cock 2.149±0.940 kg, hens 1.567±0.038 kg

Reproductive Traits: Age at first egg: 5-7 months.

Hatchability on total egg production: 60-80%, **Broodiness:** Usual

Production Traits

Egg shell colour: Light Brown, brown, dark brown, **Annual egg production (number):** 60-80, **Weight of an egg:** 46.002±1.191 g, **Dressing percentage:** 60-70%

Utility: These are used both for meat as well as egg production. In Punjab, these birds are maintained both by progressive farmers as well as by poor families. While the former keep these for home consumption, then sell live birds/chicks and eggs to earn livelihood. In Haryana, mostly these are maintained as reserves by a few low-income families located in one part of the village. These birds are also available in the slums on the outskirts of cities and the owners are doing a good business because of the readily available market.

- The population of **Punjab Brown** fowl in our country is 13,65,800 as per 20th Livestock Census (2019).

17. Mewari

- **Home Tract:** Central and Southern part of Rajasthan. The home tract of this breed is Mewari region of Rajasthan.
- The local native germplasm of Southern Rajasthan has been registered as Mewari chicken.
- Mewari breed is light to dark brown colour plumage with long arching tail feathers, single comb and yellow shank and skin.

Morphometric traits

- **Body weight:** Cock 1.9 kg, hens 1.2 kg.
- The fertility and hatchability on the fertile eggs set (FES) was 74.15±4.29% and 71.37±8.15%, respectively.
- The age at first egg (AFE) in the flock was 142 days and the age at sexual maturity (ASM) was 181.2±3.85 days. The average egg weights at 28 and 40 weeks of age were 36.61±0.29 and 42.59±0.37 gm, respectively.
- The hen day egg production (HDEP) up to 40 and 52 weeks of age was 28.93±0.13 and 59.87±0.14 eggs, respectively.
- The annual HDEP up to 72 weeks of age was 86.37± 0.13 eggs. The growth and egg production of Mewari chicken is comparable with other indigenous breeds.

Production traits

Annual egg production (numbers): 37-52, **Weight of an egg:** 53 g, **Rearing system**: Free range or scavenging system. **Utility**: Egg and meat

Fig. 38: Mewari cock and hen

- The population of **Mewari** fowl in our country is 1,14,624 as per 20th Livestock Census (2019).

18. Uttara

Distributed in Kumaon region of Uttarakhand. Plumage is black in colour and comb is single. These birds have feathered shanks which are not present in any other indigenous breed of chicken. About 18% of birds have a bunch of feathers on their head (crest/crown). Broodiness is usual. The birds are noisier and flightier. Annual egg production ranges from 125 to 160 and egg weight from 49.8 to 52.7 g. Adult weight is about 1.3 kg in cocks and 1.1 kg in hens.

Fig. 39: Uttara hen and cock

19. Haringhata Black

More than 30 years back HB, an indigenous fowl genetic resource, drew the attention of the scientists, planners and developers. Existence of HB chicken has been reported as early as 1984. The breed subsequently disappeared from the scientific community until 1994 when a draft Country Report on AnGR was prepared (Pan 1994). Later on a series of programs were undertaken on characterization and conservation of the breed during 2009 onwards (Pan, 2010, Vij *et al.*, 2015, Pan, 2016). By virtue of its superb adaptability to the backyard system, productivity and disease tolerance, farmers and farm women have shown immense interest towards rearing HB chicken. This breed is jet black in colour and a good layer can produce 130 eggs annually.

Fig. 40: Haringhata Black hen and cock

Farmers maintain the birds on scavenging with negligible supply of crop residue and kitchen waste. However, it was reported that the core breeding tract consisted of the northern part of North 24 Parganas and southern part of Nadia district of West Bengal (Vij *et al.* 2015). It had performed too well in the backward and ST populated villages of Bankura district. It was found in the On Farm Trial (OFT) result of 2018 that HB breed stood first on the survival of the fittest test in the tribal areas of Bankura district. This encouraged them to go for frontline demonstration with HB breed in two successive years, i.e. 2019 and 2020. This breed plays a significant role in the empowerment of tribal women through backyard poultry farming in West Bengal.

The population of **Haringhata Black** fowl in our country is 4,55,760 as per 20th Livestock Census (2019).

20. Aravali: (newly registered by NBAGR)

Aravali Chicken is a dual-purpose breed, distributed in Banaskantha, Sabarkantha, Aravalli and Mahisagar districts of Gujarat State. Males are having birchen plumage, while females are having shafty and/or laced plumage. These birds show excellent heat tolerance capacity and very good broodiness, hatchability and mothering ability. Average adult body weight for male is 1990 g and for females is 1618 g. Average annual egg production is 72 eggs.

Fig. 41: Aravali cock and hen

21. Frizzle Fowl

These birds are found all over the hot and humid coastal areas including Andaman and Nicobar Islands. These birds are also available on high altitude hilly tracts of North eastern states like Sikkim and adjoining areas. While at this time, the exact origins of the frizzle chicken could not be pinpointed, people believed that these birds could trace their roots to Asia, China, and the East Indies.

This frizzle character is controlled by an incomplete dominant gene. It is said that these birds have better adaptability to hot and humid climatic regions. Frizzle birds have an oval body with well-developed comb and wattles. The skin of the birds is thin and pale pink in colour. The beak and shanks have no relation with plumage. Dong *et al.* (2018) demonstrated that a deletion allele in KRT75L4 is responsible for the frizzle feather phenotype in Kirin, a Chinese indigenous chicken, and not KRT6A, as in other breeds of frizzle chickens.

The birds have single comb and ear lobes are well developed with white spots on them. The eyes are bright and well developed. Plumage colour varies among birds but the most common colours are white, brown, black and mixed colour.

Fig. 42: Frizzle Fowl

Performance characteristics

Body weight at 20 weeks of age: 1.0 kg, **Age at sexual maturity:** 185 days, **Annual egg production:** 110 numbers, **Egg weight at 40 weeks:** 53 g, **Fertility:**61 %, **Hatchability:**71 %

22. Naked Neck

The home tract of this breed is Trivandrum region of Kerala but the breed is also available through hot and humid coastal areas including Andaman and

Nicobar Island and north eastern states of India. Peoples of those regions have greater affinity for naked neck birds thanks to their better adaptability to the recent and humid climatic condition also as for better taste and flavour of meat.

The Naked Neck is a breed of chicken that is naturally devoid of feathers on its neck and vent. The breed is also called the Transylvanian Naked Neck, as well as the Turken. The name "Turken" arose from the mistaken idea that the bird was a hybrid of a chicken and the domestic turkey. Naked Necks are fairly common in Europe today, but are rare in North America and very common in South America.

The trait for a naked neck is a dominant one controlled by **incomplete dominant genes** and is fairly easy to introduce into other breeds, however these are hybrids rather than true Naked Necks, which is a breed recognized by the American Poultry Association since 1965, it was introduced in Britain in the 1920s. There are other breeds of naked necked chicken, such as the French naked neck, which is often confused with the Transylvanian, and the naked necked game fowl.

Utility: Dual-purpose breed

Morphometric Traits

Weight: Male - Standard: 3.9 kg; Bantam - 965 g; Female - Standard: 3 kg; Bantam - 850 g

Skin colour: Yellow, **Egg colour:** Light brown, **Comb type:** Single

Fig. 43: A. Naked Neck Homozygous dominant (Na/Na) Fig.43B. Naked Neck: Heterozygous (Na/na+)

The naked-neck trait which characterizes this breed is controlled by an incompletely dominant allele (Na) located near the middle of chromosome number 3. Since this allele is dominant, individuals which are either homozygous dominant (Na/Na) or heterozygous (Na/na+) will exhibit the naked-neck characteristic though the heterozygous individual will exhibit less reduction in feathering. Additionally, in tropical climates if the naked-neck trait (Na) is bred into broiler strains it has been shown to facilitate lower body temperature, increased body weight gain, better feed conversion ratios and carcass traits compared to normally feathered broilers.

23. The Bantam or Dwarf Chicken

This dwarf character is controlled by a gene (dw) is a **sex-linked recessive gene** associated with reduced body weight by about 40% and 30% in homozygous males and females, respectively compared to the normal status (DwDw, Dwdw, Dw-; Daghir, 2008). There has been a discrepancy regarding the advantage of the dw gene in heat-stressed laying hens. Its effect is more pronounced in a heavy type breed compared to a lighter one. Under heat stress conditions, introducing dwarf genes becomes more interesting for broiler breeding programs, particularly when a reduction in energy requirements and feed consumption is taken into consideration. A better adaptability of dwarf hens to heat was observed only in one experiment (Sharifi *et al.*, 2010). However, there has been a discrepancy regarding the advantage of the dw gene in heat-stressed laying hens (Wasti *et al.*, 2020).

Decuypere *et al.* (1991) concluded that the inherent heat tolerance of dw genotype in laying hens was uncertain. Under chronic heat stress, it has been found that the dw gene in fast-growing broiler chickens did not improve heat tolerance (Deeb and Cahaner, 1999; Lin *et al.*, 2006).

1. **Nicobari fowls:** These are classified as dwarf birds with short shank of 3.7 cm length at 10 weeks of age. Following are the names of the World top 10 bantam or smallest poultry breeds.
2. **Serama chicken (6 oz / 170 g):** Originating from Malaysia, the Serama chicken holds the title of the smallest chicken breed in the world. These little birds weigh no more than 7 ounces and stand at around 10 inches tall. Their upright posture and puffed-out breasts give them a distinctive appearance. Despite their size, Serama chickens are fairly decent egg layers.
3. **Dutch Bantam (20 – 21 oz / 570 – 590 g):** True bantams, Dutch Bantams were brought to Europe from the island of Java, Indonesia, in the 1600s. They have an upright carriage and an impressive long tail. Known for their ability to fly, good fencing is essential if you plan to raise them in your backyard.

4. **Japanese Bantam (21 oz / 600 g):** These are famous for its gorgeous tail, which can be taller than the actual chicken itself. While not productive in terms of egg laying, these stunning birds come in various colors and can brighten up any garden. Breeding them can be challenging due to a faulty gene, resulting in a 25% hatch failure rate1.

3. **Sebright (22 oz):** These birds are known for their laced feathers and compact size. They are delightful ornamental birds and are often kept for their unique appearance rather than practical purposes.

4. **Old English Game (24 oz):** Old English Game bantams have a long history and are prized for their fighting abilities. They come in various color varieties and are quite hardy.

5. **Barbu d'Anvers (24 oz):** Also known as the Belgian d'Anvers, these bantams have a beard and muffs. They are friendly and make excellent pets.

6. **Rosecomb (26 oz):** Rosecomb bantams have a distinctive rose-shaped comb on their heads. They are lively and active birds.

7. **Cubalaya (26 oz):** Cubalayas are small, game-like birds with a unique appearance. They have a Cuban heritage and are known for their striking plumage.

8. **D'Uccle (26.5 oz):** Belgian d'Uccles are adorable bantams with feathered legs and a sweet temperament. They come in various colour patterns.

9. **Booted Bantam (30 oz):** These birds have feathered feet and are quite hardy. They are a popular choice for backyard flocks due to their charming appearance.

Fig. 44: Dwarf chicken cock and hen

24. Scaleless

For extremely appearance, there is an autosomal recessive mutation (scaleless, sc/sc) that causes completely feather loss (featherless birds) and enhances thermotolerance, especially in fast growing broilers (Cahaner *et al.,* 2008; Renaudeau *et al.,* 2012). However, scaleless chicken might be less preferred by some societies (Besbes *et al.,* 2007; Desta *et al.,* 2013; Desta, 2021). In addition to preference, scaleless birds are susceptible to parasites, sunburn, and cold weather, as well as show difficulties in copulating (Bartels, 2003). Consequently, chickens with naked necks are preferred over scaleless ones. Scaleless (sc/sc) chickens carry a single recessive mutation that causes a lack of almost all body feathers, as well as foot scales and spurs, due to a failure of skin patterning during embryogenesis.

Fig. 45: Scaleless chicken cock and hen

25. The slow feathering phenomenon is a wonderful sex-linked mutation used for sex identification of day-old chicks in some egg-type strains at hatch, providing an accurate and low-cost effective method. **It is considered a multiple allelic trait, which has four different alleles (extremely slow, Kn, slow feathering, Ks, delayed feathering, K, and rapid feathering, k+)** switched on the same locus (McGibbon, 1977). The direct gene action concerns the growth of feathers' wings in newly hatched chicks, especially remiges and rectrices. Slow feathering strains inherently have better thermoregulation due to less heat production (Merat, 1986; Fayeye *et al.,* 2006; Sex-linked slow and fast feathering rates have been widely used in poultry breeding for **auto-sexing at hatch**. The growth pattern of wing feathers is distinctively different between the rapid and slow-feathering chicks at hatch with an accuracy of 98% (Sohn *et al.,* 2012). In chicken hybrids produced from both fast and slow-feathering genotypes, sex identification of newly hatched chicks can be carried out by personnel without training with an average of 96% accuracy (Goger *et al.,* 2017).

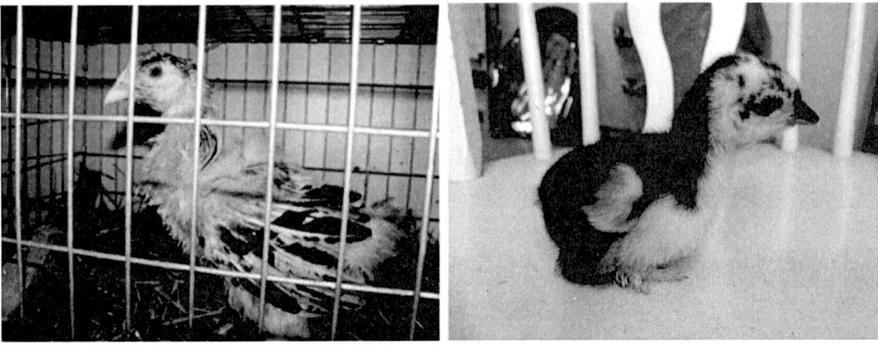

Fig. 46: Slow feathering chick

26. PD-2 (Vanaraja)

PD-2 has been developed by ICAR-Directorate of Poultry Research, Hyderabad and is used as a female parent for production of Vanaraja, **a dual-purpose backyard chicken variety**. SY-Multi Coloured broiler control population available at ICAR-DPR, Hyderabad was utilized for the development of PD-2 line. The birds have mostly single combs, multicoloured plumage, yellow colour skin and shank and lays brown coloured eggs. The average six-week body weight and shank length ranges from 450-650 g and 66-76 mm, respectively. Adult body weight at 40 weeks of age ranges from 2.4 to 2.8 kg in females and from 3 to 3.5 kg in males. The age at sexual maturity ranges from 160 to175 days. The egg weight at 40-week ranges from 52 to 56 g. The annual egg production varies between 190 and 215 eggs. Population size is approximately 5,000.

Fig. 47: PD-2(Vanaraja) cock and hen

27. Improved Varieties of Chickens

A. Gramapriya

B. Krishibro

C. Srinidhi

Fig. 48: Grampriya, Krishibro and Srinidhi breeds

- Developed by ICAR-DPR/PDP, Hyderabad, Telangana.

A. Grampriya

Purpose: Primarily an egg-laying breed.

Origin: Developed by the ICAR-Directorate of Poultry Research (DPR), Hyderabad.

Characteristics: These birds start laying eggs at around 160-165 days of age and can produce 160-170 eggs annually.

B. Krishibro

Purpose: Broiler breed, mainly raised for meat production.

Origin: Also developed by the ICAR-Directorate of Poultry Research (DPR), Hyderabad.

Characteristics: Known for their rapid growth and good meat quality.

C. Srinidhi

Purpose: Dual-purpose breed, suitable for both meat and egg production.

Origin: Developed by the ICAR-Directorate of Poultry Research (DPR), Hyderabad.

Characteristics: These birds have an optimal body weight and better egg production, making them suitable for rural poultry farming.

Japanese Quail

Scientific classification of Japanese quail

Domain: Eukaryota

Kingdom: Animalia

Phylum: Chordata

Class: Aves

Order: Galliformes

Family: Phasianidae

Genus: Coturnix

Species: *C. japonica*

Binomial name: *Coturnix japonica*

Temminck & Schlegel, (1848)

- The Japanese quail (*Coturnix japonica*), also known as the coturnix quail, is a species of Old-World quail found in East Asia. First considered a subspecies of the common quail, it is now considered as a separate species.
- Japanese quail are hardy birds that thrive in small cages and are inexpensive to keep.
- They are affected by common poultry diseases but are fairly disease resistant. Japanese quail mature in about 6-weeks and are usually in full egg production by 50 days of age. With proper care, hens should lay 200 eggs in their first year of lay.
- Life expectancy is only 2 to 2½ years.
- If the birds have not been subjected to genetic selection for bodyweight, the adult male quail will weigh about 100–140 g, while the females are slightly heavier, weighing from 120–160 g.
- **Femalse:** They have speckled chest feathers, quiet and laying eggs.
- **Males:** Plain brown chest feather, Loud call/crow at maturity, vent sexed, produces foam when pressing near vent.

Fig. 49: Japanese quail male and female

- The females are characterized by light tan feathers with black speckling on the throat and upper breast. The males have rusty brown throat and breast feathers.
- Males also have a cloacal gland, a bulbous structure on the upper edge of the vent that secretes a white, foamy material. This unique gland can be used to assess the reproductive fitness of the males.
- Japanese quail eggs are a mottled brown colour and are often covered with a light blue, chalky material. Each hen appears to lay eggs with a characteristic shell pattern or colour. Some strains lay only white eggs.
- The average egg weighs about 10 g, about 8% of the bodyweight of the quail hen.
- Young chicks weigh 6–7 g when hatched and are brownish with yellow stripes. The shells are fragile, so handle them with care.

Fig. 50: Eggs of Japanese quail

Guinea Fowl in India

The Guncari birds or Guinea fowl were developed through a selection and breeding programme from a wide base stock of indigenous Guinea fowl. These birds are improved for high disease resistance and better growth rate. Three varieties of Guncari birds viz. **Kadambari, Swetambari and Chitambari** are available for commercial utilization. It is a hardy bird, suitable for any agro-climatic condition. There is no requirement for elaborate and expensive housing. It has excellent foraging capabilities and consumes all non-conventional feed not used in chicken feeding. It is more tolerant to mycotoxins. The hard egg shell provides minimum breakage and long keeping quality. Guinea fowl meat is rich in vitamins and low cholesterol.

Scientific Classification

Domain: Eukaryota

Kingdom: Animalia

Phylum: Chordata

Class: Aves

Order: Galliformes

Family: Numididae

Genus: Numida

Linnaeus, (1766)

Species: *N. meleagris*

Binomial name: *Numida meleagris* (Linnaeus, 1758)

1. Cari-Kadambari

Production Characteristics

Body weight at 8 weeks 600-620 g, Body weight at 12 weeks 1000-1035 g

Age at first egg 230-250 day, **Egg weight** 40-43 g

Egg production (March to September) 100-120, **Fertility**-70-73 %

Hatchability on fertile eggs set-70-78 %, **Livability**-Excellent

Fig. 51: CARI-KADAMBARI breed

2. Cari-Swetambari

Production Characteristics

Body weight at 8 weeks 525-560 g,

Body weight at 12 weeks 920-960 g

Age at first egg 230-250 day,

Day Egg weight 38-40 g

Egg production (March to September) 100-115,

Fertility 70-65%

Hatchability on fertile eggs set 70-78 %,

Livability-Excellent

Fig. 52: CARI-SWETAMBARI breed

3. Cari-Chitambari

Production Characteristics

Body weight at 8 weeks 600-620 g, **Body weight at 12 weeks** 1000-1030g

Age at first egg 232-251 day, **Day Egg weight** 40-43.5 g

Egg production (March to September) 100-120, **Fertility** 70-73%

Hatchability on fertile eggs set 80-82%, **Livability**-Excellent

Fig. 53: CARI-CHITAMBARI breed

Turkey breeds

Scientific classification of domestic turkey

Domain: Eukaryota

Kingdom: Animalia

Phylum: Chordata

Class: Aves

Order: Galliformes

Family: Phasianidae

Genus: Meleagris

Species: *M. gallopavo*

Subspecies: *M. g. domesticus*

Trinomial name

Meleagris gallopavo domesticus (Linnaeus, 1758)

The domestic turkey (*Meleagris gallopavo domesticus*) is a large fowl, one of the two species in the genus Meleagris and the same species as the wild turkey. Although turkey domestication was thought to have occurred in central Mesoamerica at least 2,000 years ago, recent research suggests a possible second domestication event in the area that is now the southwestern United States between 200 BC and AD 500. However, all of the main domestic turkey varieties today descend from the turkey raised in central Mexico that was subsequently imported into Europe by the Spanish in the 16th century.

The domestic turkey is a popular form of poultry, and it is raised throughout temperate parts of the world, partially because industrialized farming has made it very cheap for the amount of meat it produces. Female domestic turkeys are called hens, and the chicks are poults or turkeyling. In Canada and the United States, male turkeys are called toms; in the United Kingdom and Ireland they are stags.

The great majority of domestic turkeys are bred to have white feathers because their pin feathers are less visible when the carcass is dressed, although brown or bronze-feathered varieties are also raised. The fleshy protuberance atop the beak is the snood, and the one attached to the underside of the beak is known as a wattle.

Fig. 54: Turkey egg

The English-language name for this species results from an early misidentification of the bird with an unrelated species which was imported to Europe through the country of Turkey. The Latin species name gallopāvō means "chicken peacock".

Fig. 55: Black Spanish turkeys

Some of the important varieties of Turkey are as follows:

- **Belts Ville White:** It is a medium sized turkey having white feathers. They produce more eggs compared to Broad Breasted Bronze and therefore included in breeding programs. The toms will be weighing 10-12 kg at maturity and hens 7-8 kg.
- **Broad Breasted Bronze:** It is the most popular and heaviest variety of turkey. As the name indicates, it has a broad and prominent chest region and bronze-coloured feathers. The males and females at maturity weigh 15-18 kg and 12-13 kg, respectively. Most of the present-day hybrid turkeys are crosses of different strains of Broad Breasted Bronze or Belts Ville White. One of the common hybrid turkeys popular in North America is Nicholas Turkey. It is the cross of the above two.

Fig. 56: Belts Ville White

Fig. 57: Broad Breasted Bronze

- **White Holland:** It is a popular variety of turkey most commonly found in European countries. It is bred and developed in Holland after importing several varieties from North America. They are also used as crosses.

Fig. 58: White Holland breed

Duck and Geese

The duck (*Anas platyrhynchos*) is one of the popular poultry species, and in India it holds a significant position in terms of egg and meat production. Though ducks are reared throughout the world, the majority are concentrated on the Asian continent. In addition to being a valuable global livestock sector generating eggs, meat, and feathers, duck farming provides small, marginal, and even landless people with a good source of income. India's duck rearing is characterised by being nomadic, extensive, seasonal, sometimes primitive, and still controlled by landless farmers, small-scale farmers, marginal farmers, and nomadic tribes.

Breeds of domestic duck

Based on their utility, ducks can be divided into three categories, i.e., egg type, meat type, and for ornamental purposes. But there are certain breeds which are very popular due to their economic traits, and the majority of the native ducks reared in India have not been categorised. Popular reared breeds of ducks are as follows:

1. Ducks reared for egg purpose are **Khaki Campbell, Bali, Indian Runner, Nageswari, Chara and Chemballi (Kuttanad ducks) etc**.
2. Ducks reared for meat purposes are **Aylesbury, Muscovy, Pekin, Rouen, China duck, Ruel Cayuga.**
3. Ducks reared for ornamental purposes are **Call, Carolina, Crested White, Mandarin, Grey calls, White Calls, Black East India breed etc.**

Among various breeds, **Khaki Campbell** is the best egg producer (Khaki Campbell duck farm at Hili Block of West Bengal) and **White Pekin** is the most popular meat duck. Khaki Campbell has a high laying capacity of 240–280 eggs/bird/year and White Pekin, being a meat-type duck, gains a weight of 2.2–2.5 kg in 40–45 days of age, an efficient FCR of 1: 2.3–2.7 kg. The National Bureau of Animal Genetic Resources under ICAR has recognised two native breeds of ducks in India, namely Maithili and Pati. The Arani ducks of Tamil Nadu are also quite popular. Certain breeds have also been developed for better performance, like Vigova Super-M, better known as the broiler duck. Other hybrid ducks include Cherry Valley (Dual purpose), Hytop (Mule duck), Legarth (Meat-type), etc.

Fig. 59: Khaki Campbell duck farm at Hili Block of West Bengal

(i) Egg type

Khaki Campbell and Indian Runner are the common egg type ducks reared in India. Their description is as follows:

- **Khaki Campbell:** This breed is developed in England by crossing Rouen, White Indian Runner and Mallard. Plumage colour is Khaki. The size of the head of male is larger than a female. Bills and shanks are black in colour. The body weight is light. Khaki Campbell duck lays 280 -300 eggs per bird per year (it can lay up to 365 eggs a year; an egg a day without a break). The standard weight of drake (male) and duck (female) is 2.2 to 2.4 kg and 2.0 to 2.2 kg, respectively.

Fig. 60: Khaki Campbell breed

- **Indian Runner:** It is next to the Khaki Campbell duck in respect of egg production and native breed of Indonesia. The three standard varieties of Indian Runner are white, penciled and fawn. The body is broader in front and slightly tapering at back. The outstanding feature of this breed is its perpendicular carriage which gives a lean appearance with wedge-shaped bill. It lays 250-280 eggs per year per bird. The standard weight of drake and duck is 1.6 to 2.2 kg and 1.4 to 2.0 kg, respectively.

Fig. 61: Indian Runner duck breed

1. Maithili duck (registered by NBAGR)

- Maithili ducks are distributed in **Motihari, Sitamarhi, Madhubani, Araria, Kishanganj and Katihar districts of Bihar.**
- These ducks have uniform light/dark brown feathers throughout the body. Circular spots on the feathers (Mosaic pattern) in ducks.
- Dark brown to ash colour in drakes. Head is bright black to greenish black in drakes and brown in ducks.

- Body carriage is slightly upright and the bill shape is horizontal.
- Average age at first egg is 191.12 days (range 159-223). Average annual egg production is 54.6 (range 33-71). Average egg weight is 49.53 g.
- Body weight at 6-month of age is 1.18 kg (range 1.12-1.24). Population size is approximately 46,000.

Fig. 62A: Female Maithili duck

Fig. 62B: Male Maithili duck

2. Pati duck: (registered by NBAGR)

- **In Assam, Pati ducks are widely distributed in Brahmaputra and Barak valley**.
- The annual egg production of Pati duck is 60 to 80 eggs.
- Growth rate of desi ducks is slower and their age at first egg is about 200-240 days.

Fig. 63: Pati duck breed

3. Andamani duck (registered by NBAGR)

It is a dual-purpose breed, distributed mainly in **Nimbudera to Diglipur region of North and Middle Andaman**. Whole body in both drake and duck is covered with black plumage with white marking under the neck extending up to the belly. Average adult body weight for drake is 1406 g and for duck is 1265 g. Average annual egg production is 266 eggs.

Fig. 64: Andamani duck breed

(ii) Meat type

Aylesbury, Muscovy and Pekin are the common meat type ducks reared in India. Their description is as follows:

- **Aylesbury:** It is a native bird of England and plumage of both sexes is white. The legs and feet are bright orange and bill is yellow in colour. This is considered a deluxe table bird because of its light bone and high percentage of creamy white flesh. The standard weight is around 4.5 kg for Drake and 4.0 kg for duck.

Fig. 65: Aylesbury duck

- **Muscovy:** It originated in South America. There are no feathers on the face and the skin is bright red in colour with caruncles around the eyes. Drake has a knob on head which gives the appearance of a crest. Voice is not characteristic of sex. The incubation period of eggs is 35 days. Muscovy's, when crossed with other breeds, produce sterile ducks called "Mule ducks". The standard weight of drake and duck is 4.5 Kg to 6.4 kg and 2.2 to 3.1 kg, respectively.

Fig. 66: Muscovy Duck

- **Pekin:** It is originated in China and its white variety is most popular for meat purposes. It has creamy white plumage, yellow flesh, long, broad and deep body with bills and legs deep orange in colour. The white Pekin attains 2.2 to 2.5 kg body weight at 7-weeks of age with a feed conversion ratio of 1:1.26-3.0 kg. It lays around 160 eggs per bird per year. The standard weight of drake and duck is 4.5 and 3.6 kg, respectively.

Fig. 67: White Peckin breed

Goose

Goose breeds are usually grouped into three weight classes: Heavy, Medium and Light. Most domestic geese are descended from the greylag goose (*Anser anser*). The Chinese and African Geese are the domestic breeds of the swan goose (*A. cygnoides*); they can be recognized by their prominent bill knob.

Kashmir Anz is a registered breed in India, Indian Council of Agricultural Research (ICAR) has recognized `Kashur Anz' as the only existing domestic goose species in the country. The Breed Registration Committee of ICAR-National Bureau of Animal Genetic Resource (ICAR-NBAGR) registered local geese of Kashmir as 'Kashur Aenz'.

The local domestic geese breed from the Kashmir Valley has now been recognised as a breed named 'Kashmir Anz', making it the first domestic geese breed listed in India. Geese rearing in the Valley dates back to ancient times. They are reared for meat and eggs and by enthusiasts in regions around freshwater bodies. Kashmir Anz geese are coloured cinnamon brown, white or a mixture, with beak colour varying from black to yellow through all intermediates. Their shanks are orange, and eyes are either grey or brown. Peculiarities like the knob, dewlap and paunch are also present in some of these geese. Two strains (or 'within breed' types) include the '**Safed Anz**' and '**Katchur Anz**'. It said that goslings of "Kashmir Anz" also have two variants, which include yellow feathered and blackish yellow feathered. "Yellow gosling appears to grow into "Safed Anz" and blackish yellow gosling into "Katchur Anz".

"Katchur Anz" are predominantly being reared in Bandipora and Baramulla, whereas Safed Anz are popular in Ganderbal. Mixed flocks of both Katchur and Safed Anz are seen in Srinagar and Budgam.

Sexual dimorphism, on the basis of plumage and eye colour, is absent. Vent sexing or vocalisation is the most accurate and practical method of gender identification. The average adult body weight of the gander is 3.82 kg and that of the goose is 3.34 kg. Body temperature, respiration and heart rates are 40.05 ± 0.15°C, 17.16 ± 0.75 breaths/min and 60.57 ± 5.09 beats/min, respectively. Each goose lays about 12 white-shelled eggs a year, weighing 137 g on average. Dressing percentage is around 68%. The acceptability of the meat is good, and a significant proportion of consumers rate it better than chicken or mutton in terms of appearance, texture, taste and overall acceptability.

Fig.68: Kashmir Anz goose breed

Their population has gone down drastically. They are now found in Wular lake of Bandipora, Ganderbal, and other districts. In Srinagar, they were found in huge numbers in Anchar side of Soura where their number has gone down. Their population is above critical level so far but requires conservation.

6

Pig Breeds of India

Pork production in India is limited, representing only 9% of the country's animal protein sources. Production is concentrated mainly in the northeastern states of the country and consists primarily of backyard and informal sector producers. The contribution of pork products in terms of value works out to 0.80% of total livestock products and 4.32% of the meat and meat products. The contribution of pigs to Indian exports is very poor.

The chromosomes of the domestic pig (*Sus scrofa domesticus*) are known to be prone to reciprocal chromosome translocations and other balanced chromosome rearrangements with concomitant fertility impairment of carriers. The domestic pig has diploid chromosome (2n) number 38. The domestic pig karyotype shows a tendency for balanced structural chromosomal rearrangements, such as reciprocal, Robertsonian, and tandem translocations, as well as inversions. There is extensive polymorphism in the heterochromatin, with G-C-rich heterochromatin located in centromere regions of biarmed chromosomes and less G-C-rich heterochromatin in the centromere regions of one-armed chromosomes.

The diploid chromosome number of the pig is 2n=38, consisting of 18 autosomal chromosome pairs, which vary in length and morphology (12-bi-armed and 6 one-armed pairs), and two sex chromosomes, XX or XY. Robertsonian translocation is known to occur in the acrocentric chromosomes 13, 14, 15, 16, 17, and 18. In these cases, the chromosomes fuse at the centromeric region, resulting in the production of two derivative chromosomes, a large bi-armed chromosome, and a secondary short chromosome often lost in subsequent cell divisions. Thus, the karyotype of a domestic pig carrying Robertsonian translocation has a distinct 2n = 37 diploid chromosome number, with the noticeable addition of a novel large bi-armed chromosome.

Importance of Pig Farming

Pigs have certain essential traits that make them valuable for farmers:

- High Fecundity: Pigs reproduce quickly.
- Better Feed Conversion Efficiency: They efficiently convert feed into meat.
- Early Maturity and Short Generation Interval: Faster economic returns.
- Pig farming also requires relatively small investments in buildings and equipment.
- It has immense potential to ensure nutritional and economic security for weaker sections of society.
- Approximately 70% of the pig population is reared under traditional smallholder, low-input, demand-driven production systems.
- Pork consumption is popular among selected populations.

Distribution of Pig Population

- Pig population is not uniform across the country.
- Thick populations of pigs are recorded in the eastern and north-eastern states.
- The highest population is in Assam, followed by Uttar Pradesh, West Bengal, Jharkhand, and Nagaland.
- Most of the pig population is concentrated in tribal belts where people are non-vegetarian.

Breeding Level

- Over 20% of the pigs kept in India are crossed with exotic breeds.
- However, inbreeding is common due to non-systematic breeding and selection.
- Pig rearing remains an unorganized venture that requires science and technology-driven support.

Table 1: Taxonomic classification of pig

Phylum	*Cordata*
Sub-phylum	*Craniata* (Vertebrata)
Class	*Mammalia*
Sub-class	*Theria* (Viviparous)
Infra-class	*Eutheria* (Placenta)
Order	*Ungulata* (hoofed mammals)
Sub-order	*Artiodactyles* (even-toed)
Sub-division	*Sunia*
Family	*Suidae* (True pigs)
Genus	*Sus*
Species	*scrofa* (European wild pigs) *domesticus* (domestic pigs) *cristatus* (Indian wild pigs) *andamanensis* (Andaman Islands) *verrucosus* (Malayan pigs) *vittatus* (Malayan pigs) *barbatus* (Malayan pigs) *salvanius* (Himalayan pigs) *africanus, procus* (African pigs)

Table 2: Registered Indian breeds of pigs (by ICAR-NBAGR, Karnal)

Sl. No.	Breed	Home Tract
1.	Nicobari	Andaman & Nicobar
2.	Doom	Assam
3.	Purnea	Bihar and Jharkhand
4.	Agonda Goan	Goa
5.	Banda	Jharkhand
6.	Manipuri black	Manipur
7.	Niang Megha	Meghalaya
8.	Wak chambil	Meghalaya
9.	Zovawk	Mizoram
10.	Tenyi Vo	Nagaland
11.	Mali	Tripura
12.	Ghurrah	Uttar Pradesh
13.	Ghungroo	West Bengal
14.	Andamani	Andaman & Nicobar

Indigenous Breeds of Pigs

1. Ghungroo

- Ghungroo an indigenous strain of pig first reported from North Bengal, is popular among the local people because of high prolificacy and ability to sustain in a low input system.

- This breed/strain produces high quality pork utilizing agricultural byproducts and kitchen wastes. Ghungroo is mostly black coloured with typical 'Bull dog' face appearance, with a litter size of 6-12 piglets, individually weighing about 1.0 kg at birth and 7.0 – 10.0 kg at weaning.
- Both sexes are very much docile and easy to handle. In the breeding tract they are maintained under a scavenging system and mainly act as insurance to the rainfed agriculture.
- At ICAR-National Research Centre on Pig, Rani, Guwahati, Ghungroo pigs are being maintained under an intensive system of rearing with standard breeding, feeding and management systems.
- Their evaluation for genetic potential for use in future breeding programmes is in progress and this indigenous strain is performing very well in terms of productive and reproductive efficiency.
- Some of the selected sows have delivered a litter size of 17 piglets at birth as compared to the other indigenous strains of pigs maintained at the Institute farm.

The population size of **Ghungroo** pigs is 2,08,751 (20th Livestock Census, 2019), which is the third highest for this breed and p**ercentage (%) share with respect to total** is 2.9.

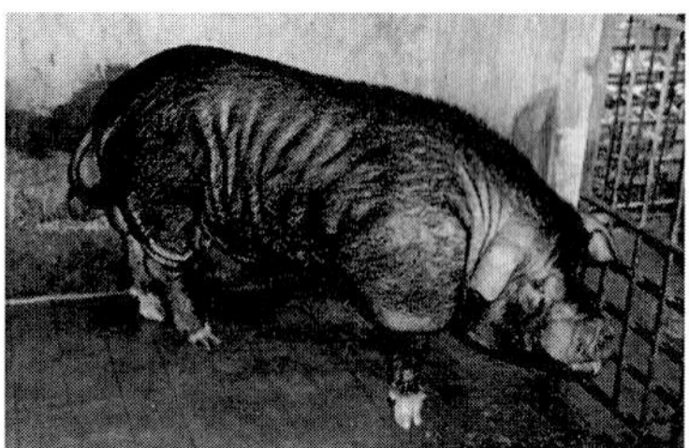

Fig. 1: Ghungroo sow and boar

2. Niang Megha

- Indigenous pigs of Meghalaya are locally known as 'Niang Megha', which are small in size with long snoot short and erect ears with black coat colour and white spot at forehead and legs. They have the ability to sustain under low input system and distributed throughout the state of Meghalaya.
- Colour: Black, star shaped white patches at forehead and sometimes hock joint.

- Visible characteristic: Long tapering snout, partially white at nostril, long bristle on midline but uniform in other places, short ears, top line- Straight in male, concave in female.
- The population size of **Niang Megha** pigs is 2,21,962 (20th Livestock Census, 2019), which is the second highest for this breed and **percentage (%) share with respect to total is 3.1.**

Fig. 2: Niang Megha sow and boar

3. Tenyi Vo

- The native Naga pig, or Tenyi vo as it is called locally, is found in the areas of the districts of Chakesang, Mao, Tuensang and Angami in Nagaland, in the extreme northeastern corner of India, along the border with Myanmar. The name tenyi vo literally translates as "pig from Angami."
- Colour: Black
- Visible characteristic: These are potbellied animals with sagging back and **pendulous belly touching the ground in females**, straight tail ending with white marking reaching the hock joint. White stocking, white markings.
- The population size of **Tenyi Vo** pig is 4,471 (20th Livestock Census, 2019) and the percentage (%) share with respect to total is 0.1 only.

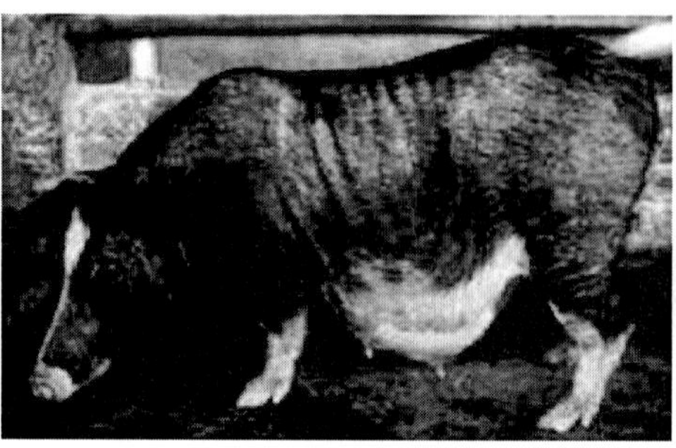
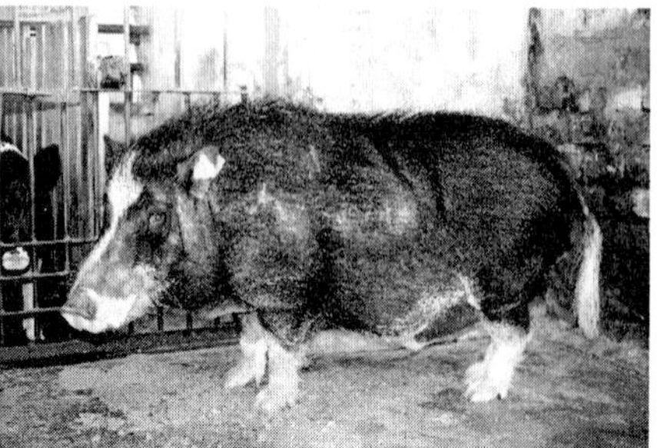

Fig. 3: Tenyi Vo sow and boar

4. Nicobari

- Synonym: Ha-un/Naut (means pig in local Nicobari language)
- **The Nicobari pig, locally known as Ha-un, is an indigenous pig germplasm located only in the Nicobar group of islands, India.**
- Nicobari pigs are semi-feral in nature and no systematic or scientific management is followed by the tribes. As such, tribes were not aware much about the farming system and the pigs were reared under a free-range system. Generally, pigs are not reared for commercial purposes.
- Colour: Black, grey, brown, blackish brown and fawn skin colour.
- Visible characteristic: Marked bristle crest(mane) on the back extending from mid head/shoulder to base of the tail, facial profile varied from flat to concave. No curling is the characteristic feature of the tail. They are fast runners.
- The population size of **Nicobari** pigs is 23,695 and the percentage (%) share with respect to total is 0.3 only (20th Livestock Census, 2019).

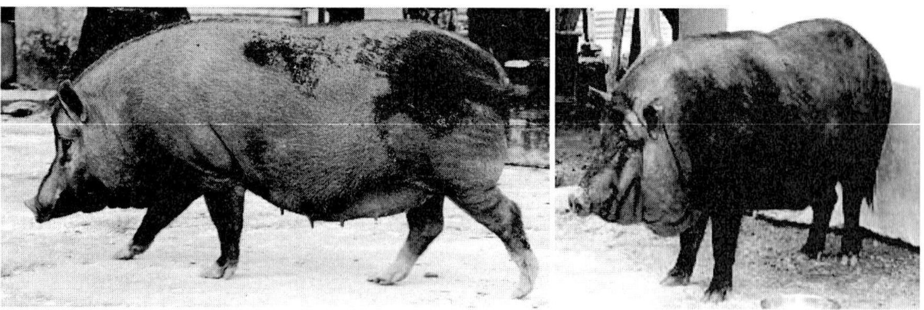

Fig. 4: Nicobari sow and boar

5. Ghurrah

- These pigs are native to Bareilly division and adjoining parts of Lucknow division of Uttar Pradesh. These are black coloured medium-sized pigs with a flat belly, angular body, and long straight snout. Legs below the hock joint are white. Thick line of hairs is present from neck to shoulders. Head is elongated with a triangular face and short leaf-shaped vertically erected ears. An adult male weighs about 46 kg and females about 48 kg. Litter size is 6.85 at birth and 5.65 at weaning.
- Colour: These are black coloured medium-sized pigs with a flat belly, angular body, and long straight snout.
- Visible characteristic: Legs below the hock joint are white. Thick line of hairs is present from neck to shoulders. Head is elongated with a triangular face and short leaf-shaped vertically erected ears.

Fig. 5: Ghurrah sow and boar

6. Mali

- Mali is native to Tripura, and is a black coloured medium sized pig with pot belly.
- Bristles are medium to small ubiquitously distributed throughout the body.
- Animals are characterized with short erect ears lying perpendicular to the body axis and concave snout.
- Adult body weight averages about 68 kg in males and 71 kg in females.
- Average litter size: 5.15 (range 3-7) at birth and 4.46 (range 3-6) at weaning. Population size is approximately 45,000 – 50,000. (Livestock Census, 2012 and NBAGR)

Fig. 6: Mali sow and boar

7. Purnea

- Purnea is a black coloured medium sized pig found in Purnea and Katihar districts of Bihar and adjoining areas of Sahibganj district of Jharkhand.
- These pigs have a compact body and pot belly. However, in few animals, white spots at the lower limbs are also seen.
- Thick line of bristle is present on topline from neck to shoulders giving the animal a wild look.
- Animals are characterized with round faces; short conical and erect ears and small, thick and slightly concave snout.
- Skin is thick with neck folds in mature animals. Adult body weight varies from 41 to 50 kg.
- Litter size at birth varies from 4 to 6. These pigs are ferocious in nature. Population size is approximately 1,00,000 – 1,20,000 (NBAGR).

Fig. 7: Purnea sow and boar

8. Doom

- Doom pig found in Assam is a newly registered indigenous breed from the subtropical ecosystem of north east India. This breed adapts well to the harsh agro-climatic condition with minimum input.

- The population of this pig breed showed a declining trend, mainly due to high rate of crossbreeding, lack of breeding policy and lacunae in the production system.
- Under field conditions, the farmers follow two different types of production systems, viz. migratory scavenging and backyard production system.
- Doom pigs reared under backyard production systems have higher growth performance and carcass yield. However, pigs in the migratory scavenging system have better reproductive efficiency and quality pork yield.
- This unique germplasm is characterized by medium body size, short ears, black coat colour with thick line of hair (bristle) on the crest extending up to lumbar region.
- The population size of Dhoom pig is 2,81,472, which is the highest in India for this breed. The and **percentage (%) share with respect to total is 3.9** (20th Livestock Census, 2019).

Fig. 8: Doom sow and boar

9. Agonda Goan

- Agonda Goan found in Goa is a local pig breed of the state traditionally known as 'Gavthi Dukor' (village pig). This breed is known for its adaptability and ability to thrive well with available feed resources.
- Agonda Goan is a small body statured, early maturing pig with rough black bristles.
- Agonda Goan breed of the pig plays a major role in the profitability of pig rearing in Goa region. Early maturity, good mothering ability, better adopt on and sustain on scavenging are the important characters of this

breed. It is better for crossbreeding with exotic breed so as to get better adoptability and disease resistance.

- Agonda Goan pigs are distributed in both the districts (North and South Goa) of Goa state and its population is sufficiently large (37,556 according to 19th Livestock census); however, South Goa seems to have more animals.
- This pig is mostly black coloured, and white patches on legs and face are found. It has rough bristles black and gray in colour with a slightly concave topline. Male animals are little wild in nature weighing 40-60 kg and get sexually mature early i.e. 78 to 120 days. Females are good mothers.
- The population size of Agonda Goan pigs is 1,325 only (20th Livestock Census, 2019).

Fig. 9: Agonda Goan sow and boar

10. Banda

- Banda breed pigs are mainly found in Ranchi, Chatra, Hazaribagh, Koderma, Giridih, Ramgarh, Bokaro, Dhanbad, Gumla, Lohardagga, Simdega, Pakur, Bokaro, Dhanbad districts.
- This small-size, black-colour, small-snout, disease-resistant and hardy breed is very active and well adaptable under agro climatic conditions of Jharkhand.
- Annual growth of pigs of this breed was observed to be 20-30 kg with litter size 3-5 in one farrowing.

Fig. 10: Banda sow and boar

11. Manipuri black

- Manipuri Black pig is native of Manipur state and as the name indicates black in colour. The Manipur Black pig is medium in size with a flat belly and short legs.
- The head of these pigs is short, slightly concave with short ears, and short to medium snout.
- White patches are sometimes seen in extremities such as legs and snout areas. Hairs are predominantly black and sparse, however few pigs with dark grey are also found.
- Bristle production is very scanty and cutting has not been practiced. Adult body weight averages about 96.0 kg in males and 93.0 kg in females.
- Average litter size is 8.27 (range 6-11) at birth and 6.02 (range 5-9) at weaning.

Fig. 11: Manipuri black sow and boar

12. Wak Chambil

- Wak Chambil is a small sized pig with a round and medium pendulous belly. They are distributed in North Garo Hills, East Garo Hills, South Garo Hills, West Garo Hills and Southwest Garo Hills of Meghalaya.
- These pigs have small heads and eyes, small erect ears, and short and pointed snouts. These pigs have thick long hair on the eyebrows and over the forehead and neck.
- Limbs are short with small hooves that partially touch the ground. Bristles are short with high density all over the body.
- Pork of this breed has unique flavour and taste; thus, it is utilized during special religious and ceremonial occasions.
- Adult body weight averages 32.0 kg in males and 29.0 kg in females. Average litter size is 5.8 (range 4 -11) at birth and 4.52 (range 3 - 8) at weaning.

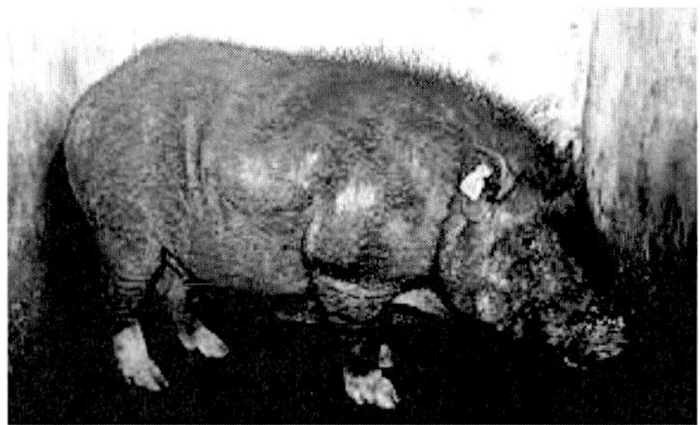

Fig. 12: Wak Chambil pig breed

13. Zovawk

- **Zovawk pigs are found in Mizoram**. These pigs are black in colour. They are small sized animals having bulging bellies.
- Synonyms: Mizo local (Literally, Zo means Mizo and vawk means pig)
- The ears are short and erect. The snout of these pigs is short and concave. The height at withers is 49.15 (45–54) cm in males and 49.04 (47–50) cm in females.
- Zovawk pigs are distributed in Mamit, Kolasib, Aizawl, Lunglei, Sahia and Champhai districts of Mizoram with an estimated population of about 40000.

- The age at first farrowing and farrowing interval is 314.9 (249–350) days and 170 (164–205) days, respectively (Breed Descriptor 2017).
- The litter size at farrowing and at weaning is 7.40±0.40 and 5.2±0.66, respectively. Carcass weight is 28.29 (23–33) kg in male animals and 34.2 (25–43) kg in female animals (Breed Descriptor 2017). The male animals of this breed also mature at very early in age and are able to impregnate the sows even before 6months of age.

Fig. 13: Zovawk sow and boar

14. Andamani

Andamani pig is a native to the islands of Andaman & Nicobar and mainly reared for meat (pork). These pigs are sturdy and medium in size. The coat is mostly black and sometimes rusty gray. They are fast runners and evolved to thrive under a low-input management system. Average adult body weight is 71 kg in male and 68 kg in females.

Fig. 14: Andamani sow and boar

Exotic Breeds

1. Large White Yorkshire

- The Large White breed of pig was developed in England in the late 1700s. There were nearly 4000 Large Whites registered in England in 1981, **ranking them as the top breed in that country. It is also the leading breed of the world**, as Yorkshires in the USA and Canada are direct descendants of the Large White.
- **Known as "The Mother Breed"**
- Entirely white in colour, but they have **freckles** (black pigment spots).
- Large White pigs possess good carcass quality. The breed is highly prolific and an efficient feed utilizer.
- **Most extensively used exotic breed in India**
- Body colour is solid white with occasional black – pigmented spots
- Erect ears, snout of medium lengths and dished face
- **Excellent breed for the purpose of cross breeding**
- Prolific breeds
- Mature boar 300-400 kg
- Mature sow 230-320 kg

Fig. 15: Large White Yorkshire breed of pig sow and boar

2. Landrace

- The Landrace breed was developed in Denmark by crossing the native pig with the Large White. This cross was then improved on during years of testing and breeding under strict government control. The Danes refused to export live pigs until World War II, when the best specimens of the breed were exported to Sweden.

- This breed is also white in colour with freckles. It is known to have long, deep sides; relatively short legs; square ham; heavy lop ears; trim jowl.
- It has leaner carcass than meat as well as less back fat and lard. Landrace breed is highly prolific and an efficient feed utilizer.

Fig. 16: Landrace breed of pig sow and boar

3. Tamworth (Irish grazer) or Golden brown (or) golden red

- The Tamworth originated in Ireland where they were called "The Irish Grazer". About the year 1812 it is said that Sir Robert Peel, being impressed with the characteristics of them, imported some of them and started to breed them on his estate at Tamworth, England.
- A long legged, lean pig.
- Sows weigh around 260 kg and boars, 320 kg.
- Their coat is a distinctive ginger coat.
- They have prick ears, a long snout and alert expression.

Fig. 17: Tamworth breed of pig sow and boar

4. Berkshire

- Berkshire, a breed of domestic pig originating in England, where in the early 19th century the name "Berkshire" became synonymous with improved pig strains of differing origin and type.

- Black coat with 6 white points, one each on 4 feet, on the face and on the tip of the tail.
- The Berkshire is medium-sized and predominantly black in colour, with white on its face, legs, and tip of tail.
- It has a short-dished face with erect ears pointing slightly forward. The breed is used for fresh pork production in England, Japan, North and South America, and other areas worldwide.

Fig. 18: Berkshire sow and boar

5. Duroc

- The Duroc pig is an older breed of domestic pig. The breed was developed in the United States and formed the basis for many mixed-breed commercial hogs.
- Well-known to have excellent weight gain rate and as a high feed converter. The sow matures early, produces large litter (up to 15 piglets) and fair mothering ability. Have good carcass quality.
- The Duroc is a large-framed, late-maturing type, excellent for heavy-carcass production. The forequarters, particularly the head and neck, are light and it has small lop ears.
- The skin is a solid reddish colour, varying from gold to a deep, brick red. They have a medium length and slight dish of the face.

Fig. 19: Duroc breed of pig sow and boar

6. Poland China

- Oldest American breed of swine; prolific (produce 16 – 17 piglets per farrow) and good meat and carcass quality.
- Characteristics: The Poland China usually displays the coloration of the Berkshire: solid black, with white points on the nose, tail and feet. It is a large pig, heavy-jowled, lop-eared and short-legged. It is among the heaviest of pig breeds: sows average some 240 kg, boars about 50 kg more.

Fig. 20: Poland China sow and boar

7. Hampshire

- Erect ears, black body and a white band around the middle, covering the front legs.
- Hampshire hogs are black with a white belt. **They have erect ears. The belt is a strip of white across the shoulders that covers the front legs around the body.**
- The Hampshire, which is a heavily muscled, lean meat breed, is the fourth most recorded breed of the pigs in the United States.

Fig. 21: Hampshire sow and boar

8. Middle White Yorkshire

- The Middle White pig is a rare breed of domestic pig which is native to the United Kingdom.
- It originated in Yorkshire about the same time as the Large White breed.
- Its size is between the now-extinct Small White and the Large White pig, which is named for its size.
- The breed was actually and fully recognized as a breed in 1884.
- Instead of bacon or lard type pigs, the Middle White pig is known as a pork producer.
- As the name suggests, the Middle White pig is a medium sized breed of pig. It has many characteristics same to the general characteristics of Large White pigs.
- But the Large White pigs have much more dished face and they are slightly larger than the Medium White pig breed.
- It is much faster grower than its large relative.
- The animals are mainly white coloured and it has a well-balanced and meaty body.
- The animals on average reach around 61.8 kg within their 17 weeks of age.
- The Middle White pigs are much hardy animals. They are found scattered over England today. And generally bred straight in England and have not captured much of a share of the commercial pig market.

Fig. 22: Middle White Yorkshire sow and boar

9. Wessex Saddleback

- The Wessex Saddleback or Wessex Pig is a breed of domestic pig originating in the West Country of England, (Wessex), especially in Wiltshire and the New Forest area of Hampshire.
- **It is black, with white forequarters**. In Britain it was amalgamated with the Essex pig to form the British Saddleback, and it is extinct as a separate breed in Britain.
- However, the Wessex Saddleback survives in Australia and New Zealand.
- The Wessex Saddleback is black, with a white band about the forepart of the trunk, extending from one fore-foot over the shoulder to the other, forming a white band resembling a saddle (or "sheet"). It is a tall, rangy animal, adapted to foraging in woodland, its traditional use.
- The **Hampshire pig has erect ears, while the Wessex Saddleback has lop ears.**

Fig. 23: Wessex Saddleback sow and boar

7

Camel Breeds of India

Camels could be exploited for augmenting the country's milk production capacity as a supplementary milk resource in India for the benefit of increasing human population. Since milk production potential of camels in India have remained unexploited and considering its therapeutic utility it can fetch good prices. Camels are raised for milk, meat, fibre (wool and hair), transport and other work; their dung is used as fuel. Milk is often the most important camel product and is the staple food of nomads.

The camel is an important species of the Indian fragile desert ecosystem, a proven icon of adaptation with its unique bio-physiological characteristics and has formidable ways of living in harsh situations of arid and semi-arid regions. The proverbial **Ship of Desert** earned its epithet on account of its indispensability as a mode of transportation and draught ability in desert. However, its utilities are subject to continuous social and economic changes. The camel has also played a significant role in civil law and order, defence and battles from the ancient times till date. Camels formed an important component of the Mauryan Army (C.322-232 BC) and continued through the Mughal period (1200-1700 AD) to the present times.

Camels are found in Africa and Asia and are kept mostly by nomads. There are two species of camels: one-humped Arabian camels or dromedaries (*Camelus dromedarius*), the camels of the plains; and two-humped Bactrian camels (*Camelus bactrianus*), the camels of the mountains. Camels are raised for milk, meat, fibre (wool and hair), transport and other work; their dung is used as fuel. Milk is often the most important camel product and is the staple food of nomads.

There is growing recognition of the value and benefits of camels for their milk, meat and fibre. Camel dairy products could not only provide more food for people in arid and semi-arid areas but also give nomadic herders a rich source of income.

As camel milk is normally produced under low-input, low-output systems, five litres a day is considered a decent yield. Lactating camels generally produce between 1000 and 2700 litres per lactation in Africa, but camels in South Asia were reported to produce up to 12000 litres per lactation. Camels reach the maximum yield in the second or third month of lactation and produce milk for between eight and eighteen months. The daily milk yield during the wet season is often twice that of the dry season. The lactation curve of dairy camels is similar to that of dairy cattle, but camels have more persistent lactation. Arabian camels generally have a much higher milk yield than Bactrian camels and are being used increasingly in intensive dairy operations (FAO). Camel milk is unique in terms of having low fat (1.5-3.0%), low protein (2.5%) has longer shelf life, higher ratio of ß-casein to k-casein, absence of Lysozyme C and ß-lactoglobulin and presence of whey acidic protein (WAP) and peptidoglycan recognition protein (PGRP).

Did you know?

- Camels are kept for milk production in Africa and Asia.
- The global camel population is estimated at almost 39 million head.
- In sub-Saharan Africa, camels contribute about 8 percent of total milk production.
- Kenya is the largest camel milk producer in the world, followed by Somalia and Mali (FAO).

Scientific Classification of Camel

In this section, we will learn about various features of camels such as Kingdom, Phylum, Class, Order and Genus. The camel species has chromosome number 74 (2n). The classification of the Camel is given below:

Table 1: Classification of the Camel

Kingdom	Animalia
Phylum	Chordata
Class	Mammalia
Order	Artiodactyla
Family	*Camelidae*
Genus	*Camelus*

Types of Camels

There are three existing camel species living in this world currently.

1. Bactrian Camel (*Camelus bactrianus*)

Also known as the **Mongolian camel** or domestic Bactrian camel, the Bactrian camel is a **large, even-toed ungulate native to the Central Asian steppes**. Their name is derived from the **historical ancient region of Bactria.**

- In comparison to the single-humped dromedary camel, it has **two humps on its back**.
- Its two million population exists predominantly in the domestic form.
- Since ancient times, domesticated Bactrian camels have acted as pack animals in Inner Asia.
- Bactrian camels are resistant to cold, drought, and high altitudes.
- Bactrian camels have a remarkable ability to go without water sometimes for months at a time, but they can drink up to 57 litres at once when water is available.
- Bactrian camels are also said to be **great swimmers**.
- The sense of sight is well-formed and Bactrian camels have an exceptionally strong sense of smell.
- It is estimated that the average Bactrian camel **lifespan is up to 50 years, sometimes 20 to 40** years in captivity.
- **The Bactrian camel is the largest living camel.**
- The height of the shoulders is 180 to 230 cm (5.9 to 7.5 ft), the length of the head and neck is 225-350 cm (7.38-11.48 ft), and the length of the tail is 35-55 cm (14-22 in). The average height at the top of the humps is 213 cm (6.99 ft).
- Bactrian camel weight varies between 300 and 1,000 kg, and males are often much larger and heavier than females.
- The colour of its long, woolly coat ranges from dark brown to sandy beige. There is a long-hair mane and beard on the neck and throat, with hair up to 25 cm (9.8 in) long.

2. Arabian Camel or Dromedary (*Camelus dromedarius*)

Also known as the Arabian camel or desert camel, the dromedary is a large, even-toed ungulate of the Camelus genus, with **single hump** on its back.

- It is the tallest of the three camel species, adult male camel height is between 1.8-2.0 m (5.9-6.6 ft), while female camel height is 1.7-1.9 m (5.6-6.2 ft).

- Typically, males weigh 400 to 600 kg, and females weigh 300 to 540 kg.
- Its long, curved body, narrow chest, single hump, and long hair on the throat, shoulders, and hump are distinctive features of Dromedary species.
- The coat is a shade of brown in general.
- The hump is made of fat bound together by fibrous tissue, 20 cm high or more.
- During daylight hours, dromedaries are primarily active.
- They form herds of approximately 20 males, led by a dominant male.
- This camel feeds on foliage and desert plants, allowing it to survive in its desert habitat through many adaptations, such as the ability to withstand losing more than 30% of its total water content.
- In the semi-arid to arid regions of the Old World, mostly in Africa and the Arabian Peninsula, the domesticated dromedary is usually found, and a large feral population exists in Australia.

3. Wild Bactrian Camel (*Camelus Ferus*)

- A critically endangered species of camel living in parts of northwestern China and southwestern Mongolia is the wild Bactrian camel.
- It is closely related to the camel of Bactria (***Camelus bactrianus***). They are both big, double-humped, even-toed ungulates native to the Central Asian steppes.
- Today, only around 1,000 camels survive. Most live in China's Lop Nur Wild Camel National Nature Reserve, and a smaller population lives in Mongolia's Strictly Protected Great Gobi Forest.

Fig. 1: Wild Bactrian Camel (*Camelus Ferus*)

Table 2: Camel breeds of India registered to NBAGR

Sl. No.	Breed	Home Tract
1.	Kharai	Gujarat
2.	Kutchi	Gujarat
3.	Malvi	Madhya Pradesh
4.	Bikaneri	Rajasthan
5.	Jaisalmeri	Rajasthan
6.	Jalori	Rajasthan
7.	Marwari	Rajasthan
8.	Mewari	Rajasthan
9.	Mewati	Rajasthan and Haryana

1. Bikaneri

Bikaneri breed of camel is one of the major camel breeds found in India. The breed derives its name from the city Bikaner which was established by Rao Bika in the 15th century and is known for better draught potential.

Habitat and Distribution: Bikaneri camels are predominantly bred in Bikaner and nearby districts, such as Sri Ganganagar, Hanumangarh, Churu, Jhunjhunu, Sikar and Nagaur of Rajasthan and adjoining parts of Haryana and Punjab state. The breeding tract extends in the east from 71°53' to 78°15' longitude and in the north from 24°37' to 30°30' latitude. The home tract of this breed is arid and sandy with extreme hot and cold climates.

Features

- The camels of Bikaneri breed are heavily built and are attractive with a noble look. It has good height, a strong build and active habits.
- The colour of the coat varies from brown to black, however in some animals' reddish tinge is also found.
- They have a symmetrical body and slightly dome shaped head. The forehead has a well-marked depression (stop) above the eyes, which is characteristic of this breed. Nose is long and extends up to two third of the head. **Some camels of this breed have a luxuriant growth of hair on their eyebrows, eyelids and ears, they are called 'jheepra'**.
- The chest pad is well developed and placed between angles of elbow. The shoulders are strong, broad and well set to the chest.
- Neck is thick, fairly erect, with a marked curve giving a graceful carriage to the head. The udder is well developed in females.

Fig. 2: Bikaneri camel

- The population of **Bikaneri camels** in our country is 74,551 as per 20th Livestock Census, 2019. This breed shares 29.6% among total and first in India.

2. Jaisalmeri

Habitat and Distribution: The breeding tract of Jaisalmeri breed encompasses the **Jaisalmer, Barmer and part of Jodhpur district in Rajasthan.** The breeding tract extends in the east from 69°30' to 73°04' longitude and in the north from 24°37' to 28°15' latitude with very poor vegetation. Sand dunes are the typical features of the tract.

Features

- The Jaisalmeri camels are of active temperament and are quite tall with long and thin legs.
- They have small heads and mouths with narrow muzzles.
- The head is well carried on a thin neck and the eyes are prominent.
- The forehead is not dome shaped and is without any depression above eyes (stop).
- Also, there is no luxuriant growth of hairs on their eyebrows, eyelids and ears. The body colour is predominantly light brown.
- The Jaisalmeri camels have thin skin and short hairs on their body. The udder is mostly round in shape. It is a medium sized breed of camel.

Fig. 3: Jaisalmeri camel

- The population of **Jaisalmeri camels** in our country is 47,975 as per 20th Livestock Census (2019). This breed shares 19% among total and first in India.

3. Mewari

The Mewari breed of camel has derived its name from the **Mewar area of Rajasthan** and is well known for milk production potential.

Habitat and Distribution

The major breeding tract of the breed encompasses the **Udaipur, Chittorgarh, Rajsamand districts and adjoining Neemuch and Mandsaur districts of Madhya Pradesh**. The camels of this breed can also be seen in **Bhilwara, Banswara, Dungarpur districts and Hadoti region of Rajasthan**, which can be considered as a minor breeding tract of the breed. The breeding tract extends in the east from 73°02’ to 77°20’ longitude and in the north from 22°55’ to 25°46’ latitude with fairly good vegetation and rainfall. Average height from the main sea level is about 575 meters. **The tract consists of hills of the Aravali in the Mewar area.**

Features

- Genetic improvement of indigenous camels through conventional and molecular techniques.
- Mewari camels are stouter and a little shorter than Bikaneri.
- They have strong hindquarters, heavy legs, hard and thick foot pads.

- Well adapted to travel and carry loads across hills. The body hairs are coarse, which protects them from the bites of wild honeybees and insects.
- The body colour varies from light brown to dark brown but some animals are almost white in colour, such variation in body colour is generally not seen in other breeds of camel. The head is heavy, set on a thick neck.
- Unlike the Bikaneri camel, the Mewari camel has no 'stop', but its muzzle is loose. Ears are thick and short, set well apart, and the tail is long and thick. The milk vein is prominent and the udder is well developed in females.

Fig. 4: Mewari camel

4. Jalori

Local Name/Synonyms: **Jalori, Sanchori**

Background for such name: Named after the habitat, Jalore

Communities responsible for breeding: Dewasi are the traditional camel breeders

Geographical distribution and breeding tract

The Jalori camel derives its name from the place of rearing. The geographical distribution of the breed encompasses chiefly the **Jalore and Sirohi** districts of Rajasthan. The overall population of Jalori camels in the breeding tract was estimated to be 7906 heads.

Breed Characteristics

- Body colour - The predominant colour of Jalori camels is brown. However, it varies from light brown to dark brown.
- Head Profile- The head in Jalori camel is medium in size and is well carried on a thin neck.
- The eyes are prominent. Unlike Bikaneri camels, in Jalori camels, the forehead is not dome shaped and has no "stop", which is a name given to a depression on the frontal bone at the upper edge meeting the parietal bone.
- The supraorbital foramen, which is in the form of a deep fissure at the rostro-medial margin of the orbit, is normal in depth as compared to the Bikaneri camel where it is deep.
- The muzzle is narrow and mostly pointed in camels of Jalore district but rounded in the camels of Sirohi district.
- **Ears are up-right and set well apart. The typical adaptive feature of desert camels, the "Jheepra", is absent in Jalori camels. The lower lip is not droopy as seen in Kachchhi camels.**

Body and Stature

- It is a medium sized breed of camel.
- The Jalori camels are of active temperament.
- The neck and legs are thin.
- The body hairs are coarse in quality and medium in length.
- The chest pad is well developed.

Udder Characteristics- The milk vein is small to medium in size. The udder is mostly round in shape. Each udder quarter has a small cone shaped teat with two canals in it. The Jalori Breed is a multipurpose breed of camel and the animals of this breed are being **utilized for milk production, tourism, riding and safari.**

Fig. 5: Jalori camel

5. Mewati

Breeding tract

Alwar, Bharatpur (Rajasthan)

Main use:

Work – Riding, baggage, draught.

Visible characteristics

These camels are hard footed and slightly shorter in stature than the Bikaneri

Fig. 6: Mewati camel

6. Kachchhi / Kutchi

Habitat and Distribution

The Kachchhi breed inhabits the Ran of Kachchh in Gujarat state. The major breeding tract encompasses **the Kachchh and Banaskantha districts of Gujarat and** it extends in east from 68°20' to 74° longitude and in north from 22°51' to 24°37' latitude. The land is marshy with abundant salt bushes.

Features

- The camels of this breed are generally brown to dark brown in colour with absence of hair on eyelids and ears. The body hairs are coarse.
- Head is of medium size without a distinct "stop". Body size is medium. Camels of this breed are heavy and dull in appearance.
- They are stouter and a little shorter.
- They have strong hindquarters, heavy legs, hard and thick foot pads and are well adapted to the humid climate and marshy land of Kachchh.
- In some animals the lower lip is droopy due to which the teeth are visible from a distance. The udder is well developed and mostly round in shape.
- The population of **Kachchhi camels** in our country is 16,905 as per 20th Livestock Census (2019).

Fig. 7: Kachchhi / Kutchi camel

7. Malvi

- Malvi camel is a very specific breed of northern Madhya Pradesh (India), which has great potential for milk production. However, the Malvi camel has long been recognized as a distinct population by local camel breeders.

- Breeding tract: Mandsaur (Madhya Pradesh)
- Main use: field work, transport; food purpose for milk.
- Comment on main use males are used mostly for carrying loads, while females are reared for milk.
- Origin:
- The Malvi camel is named after the name of its breeding tract i.e., Malwa region.
- Colour, off–white
- Visible Characteristics: The most typical external characteristic of the Malvi camel is its very light or off-white colour. Malvi breeding herds show virtually no colour variation. Another typical phenomenon is its small body size and is probably the smallest of all India breeds.

Fig. 8: Malvi camel

8. Marwari

The Marwari camel derives its name from the Marwar region. The ruling family of Marwar, the biggest state of Rajputana, maintained a camel corps which was deployed in their frequent military campaigns, and on occasion it was lent out to the Mughal emperors in Delhi.

Marwari camels are found in the Marwar area specifically in Jodhpur, Jalore, Barmer districts of Rajasthan. Marwar (also called Jodhpur region) is a region of south western Rajasthan in north-western India. It lies partly in the Thar desert. The region includes the present-day districts of Barmer, Jalore, Jodhpur, Nagaur, Pali and parts of Sikar. Distribution of the breed is now mainly confined to Pali, part of Jodhpur and Barmer.

- They have heavy builds with long legs, muscular bodies and are good for agricultural operations. These animals resemble Bikaneri in several body confirmations **except facial characteristics which are at variance.**
- The physical appearance of any animal is chiefly defined by the body colour followed by stature and other phenotypic characteristics. The predominant colour of Marwari camels is brown. However, it varies from sand brown to dark brown.
- The head in Marwari camel is small to medium in size and well carried on a thin neck. Forehead is normal /flat.
- The Marwari camels have medium to large body size. They have an active temperament. The hump is medium in size. Chest pads are well developed. The body hairs are coarse in quality and medium in length.
- Udder Characteristics: The animals have good milk yield potential. Round as well as pendulous udder is found in the Marwari camel.

Fig. 9: Marwari camel

- The population of **Marwari camels** in our country is 2,720 as per 20th Livestock Census (2019).

9. Kharai

- Kutch's unique breed of camels that **can swim in seawater**, known as Kharai due to its habitat and eating habits, has been identified as a separate camel by the National Bureau of Animal Genetic Resources (NBAGR) based in Karnal, Haryana.
- Recognized as a breed, Kharai has now become the ninth breed of camel found in India.
- Kharai camels have a special ability to swim in seawater and feed on saline plants and mangroves, which is how they get their name, **Kharai**

('salty' in Gujarati). They are also known as dariyataru (meaning sea-swimmer).

Fig. 10: Kharai camel

Habitat: Kutch, a coastal region of Gujarat, which is also a large desert land, has two camel breeds. One is the popular Kutchi breed and the other, the Kharai breed, native to the region.

- The Kharai breed has the special ability to survive on both dry land and in the sea, making it an ecotonal breed.
- Recognized as a separate breed a few years ago by the National Bureau of Animal Genetic Resources(NBAGR), the Kharai camel is probably the only domesticated breed of camel that lives in dual ecosystems.
- NBAGR also certified the breed as ninth camel breed found in India, separating it from Kutchi camel.

Feeding habits: Kharai camels are known to feed on mangroves on the island offshore and to eat this salty marine food, they sometimes swim for hours.

Unique features

- These camels have a special ability to swim in seawater and feed on saline plants and mangroves, which is how they get their name, Kharai ('salty' in Gujarati).
- Their gently padded hooves help them navigate the wet and salty coastal land with ease and they can swim up to three kilometres (1.8 miles)
- During the rainy season, they swim along the Gulf of Kutch, an inlet of the Arabian Sea, to small forest islands and graze on mangroves and other saline-loving plants.
- They are also known as dariyataru (meaning sea-swimmer).

- They have adapted to the extreme climate of the desert, shallow or deep-sea waters, and high salinity.

Population

As per latest counting, the state has 6,200 camels out of which around 2,200 are found in areas such as Lakhpat, Abdasa, Mundra and Bhachau in Kutch, whereas the remaining are seen in South Gujarat near Aliya Bet. The population of Kharai camels in our country is 4,266 as per 20th Livestock Census (2019).

Threats

- Industries in Kutch–salt, thermal power, cement and shipyards, among others–pose a huge threat to the dwindling mangroves.
- Most of these industries require constructing jetties in the sea, which results in the cutting down of mangroves that are fodder for the Kharai camels.
- The increase in salinity throughout the region and the growth of industrial activities has minimized the availability of camel food and water sources

Conservation status: International Union for Conservation of Nature (IUCN)

IUCN: Endangered

Double hump camel

A very small number of double-humped camels are available in Ladakh area.

Camelus dromedrarius- One humped/Arabian/ Dromedary

Camelus bactrianius–Two humped/Bactrian camel

- India stands third after Somalia and Sudan.
- The total Camels contribute around 0.08% of the livestock population.
- 0.4 million numbers (Census, 2012)
- The chromosome number is 74.
- Two humped camels inhabit in the desert of central Asia reaching up to Magnolia and Western part of China.
- Single hump Arabian camels are widespread throughout middle east India and north Africa.
- Union Minister of Agriculture and Farmers Welfare, said the number of

camels has come down from 4 lakh (Livestock Census, 2012) to 2.52 lakh (Livestock Census, 2019), showing a decline of 37%.

Fig. 11: Double hump camel

- A small population (450 on 20th livestock census) of Bactrian camels exists in the Nubra Valley of Ladakh (Jammu and Kashmir).

Properties

- The humps are thick and flexible. At the end of winter when pastures are depleted, the humps collapse.
- The skull bone is relatively shorter and broader than that of the dromedary camel.
- The body of a camel is short and thick.
- The colour of the body varies from light brown to dark brown.
- Long hair grows on the top of the head, lower part of the neck, hump and legs.
- Adult body weight ranges from 450 to 550 kg. Male animals are heavier than females.

Fig. 12: Two humped/Bactrian camel

Origin: It is believed that these camels belong to the stock originally native to Gobi Desert of Asia and introduced by traders of Yak through silk route.

Major utility: Work - Transport.; Wool - hair

Comments on utility: These camels were widely used as an important source of transport when the silk route was in operation before 1950. They produce superior quality hair which is used in cottage industries.

Comments on breeding tract: Available in Nubra valley of Ladakh (Jammu and Kashmir).

Adaptability to cnvironmcnt: Thcsc camcls arc suitable for cold desert areas.

13 Aug 2021 — Since the 1960s, the number of Bactrian camels has grown from a mere handful to around 300 now; it can be expected to increase further.

8

Horse and Pony Breeds of India

In some developing countries, milk from horses (*Equus caballus*) and donkeys (*Equus asinus*) is an essential food for poor farmers. Generally, horses are more commonly used for dairy purposes in cool areas and donkeys in dry semi-arid regions. Milking equines is time-consuming and has to be repeated five or six times a day. In addition, an equine will not release milk unless it is stimulated by the presence of its foal. Mare's milk is commonly consumed in the steppe areas of Central Asia, where a traditional **lactic-alcoholic beverage called koumiss is produced through fermentation.** Horse milk is also an important source of animal protein for pastoralists in Mongolia. The consumption of donkey milk has become very marginal. In some African communities, donkey milk is consumed for medical purposes.

Equus is a genus of mammals in the family *Equidae*, which includes horses, asses, and zebras. Within the *Equidae*, *Equus* is the only recognized existing genus, comprising seven living species. Like *Equidae* more broadly, *Equus* has numerous extinct species recognized only from fossils. Equines are odd-toed ungulates with slender legs, long heads, relatively long necks, manes (erect in most subspecies), and long tails. All species are herbivorous, and mostly grazers, with simpler digestive systems than ruminants but able to subsist on lower-quality vegetation. The chromosome number of a horse is 64 (2n).

Following are the existing species in the genus *Equus*:

1. *E. africanus* (African wild ass)
2. *E. caballus* (domestic horse)
3. *E. ferus* (wild horse)
4. *E. grevyi* (Grevys zebra)
5. *E. hemionus* (onager)
6. *E. kiang* (kiang)
7. *E. quagga* (plains zebra)
8. *E. zebra* (mountain zebra)

- **India has eight registered breeds of Horse and Pony (NBAGR).**

Table 1: List of horse breeds in India

Sl. No.	Breed	Home Tract
1	Bhutia	Sikkim and Arunachal Pradesh
2	Kathiawari	Gujarat
3	Manipuri	Manipur
4	Marwari	Rajasthan
5	Spiti	Himachal Pradesh
6	Zanskari	Jammu and Kashmir
7	Kachchhi-Sindhi	Gujarat and Rajasthan
8	Bhimthadi	Maharashtra

Description of Horse Breeds

1. Bhutia

- The Bhutia horse, found in the Himalayan regions of Nepal, India, and Bhutan, is a breed of small and compact horses bearing resemblance with the Tibetan and Mongolian equine breeds.
- It is suited to the rough terrain and cold climate of the mountains, where it is useful for transportation work.

History and Development

- The ancestry of these horses is unclear, though today's Bhutia Horses are believed to have evolved because of extensive interbreeding between local horses and ponies.
- Crossbreeding among Bhutia, Spiti, and Tibetan ponies for many years have led to the loss of their individual characteristics.
- Therefore, these breeds are often collectively named as Indian-Country-Bred ponies.
- Furthermore, lack of organized breeding programs and shortage of fodder have affected the size and strength of Bhutia Horses.
- Population is very small, only 2,340 as per 20th Livestock Census, (2019).

Fig. 1: Bhutia horse breed

2. Kathiawari

- Kathiawari is a sturdy breed of Indian horses hailing from the Kathiawar peninsula. It has close resemblance with the Marwari and the Arabian horses, though the Kathiawari is slightly stockier and taller than the finely featured Marwari horse.
- These horses were originally bred for use as cavalry mounts and warhorses that could travel long distances on very little rations.

History and Development

- It is believed that horses inhabited the Indian west coast long before the Mughal Empire was established. During the Mughal Empire, Arabian horses were brought to the Indian subcontinent and bred with local horses, which created the ancestors of today's Kathiawaris. The addition of Mongolian horse blood also improved the breed. This breeding practice was continued even during the British Rule in India.
- Traditionally, noble households selectively bred sleek, wiry horses that could tolerate harsh conditions and carry people with weapons for long periods, and still be quick and light in action. This tradition of horse breeding among the nobility ended when India became independent.
- Apart from the Kathiawar peninsula, these horses are now found in the Indian states of Rajasthan and Maharashtra. The government of Gujarat maintains studs at several districts including Junagadh, Amreli, Porbandar, Surendranagar, and Rajkot, some of which aim towards preserving the horse breed while others look to improve the local stock. A breed registry is maintained by the Kathiawari Horse Breeders' Association.

Interesting Facts

- The Kathiawari were known for their courage and loyalty in battle, defending their master even after being wounded.
- Some of the most beautiful Kathiawari horses are now produced at the Panchal area of Gujarat.
- The most distinguishing characteristic of these horses is their inward-curving ears, which touch or may overlap each other at the tips.
- Population is 25,280 as per 20th Livestock Census (2019). The percentage (%) share with respect to the total is 7.4, which is the second highest in India.

Fig. 2: Kathiawari horse breed

3. Manipuri

The Manipuri pony is a rare Indian equine breed popular as a polo and military horse. Its striking features include its hardy stature, high endurance level, and versatile nature.

History

There have been varied opinions and contradictions regarding the source of origin of the Manipuri horse. Some experts opine that the Tibetan Pony, imported to India, over 1000 years ago, is the ancestor of the Manipuri pony. Others believe it to be developed by **crossing the Arabian and Mongolian Wild Horse**. Being bred for more than 100 years in Manipur (northeastern part of India), the Manipuri Pony was first mentioned in the Manipur Royal Chronicle (1584). The next century witnessed its frequent reference in Manipuri literature.

The Manipuri pony was immensely sought after by the British colonists from 1859 to 1916 as mounts to play a game of polo, which had been immensely popular in Manipur as early as the seventh century.

Present status

Their population has gradually declined from 2000 to 1000 ponies on an average. The reason for this might be that these horses were smuggled to Burma in massive amounts where their demand seems to be high. For the purpose of promoting them as well as preventing this breed from becoming extinct the Manipuri Horse Riding and Polo Association formed a heritage park in the year 2005.

According to the livestock census conducted in the year 2012, their numbers were as low as 1100, and might soon be declared as endangered species.

Interesting Facts

In the present time also, they are used for a particular variety of polo named the "sagol kangjei" generally played with seven players in one side, where these ponies are used throughout the game.

Population is 3,552 as per 20th Livestock Census (2019). The percentage (%) share with respect to the total is 1.0.

Fig. 3: Manipuri horse breed

4. Marwari

The Marwari horse is a breed of royal historic horses that were created in India and showed their skills as war horses for centuries. Famous for their beautiful features and ambling gaits, these are equines that trot with great style and lightness. The horses are also known for their hardiness, and are genetically similar to the Kathiawari breed originating from the same region. Presently, pure specimens of the Marwari horses are quite rare, while the government of India has taken necessary steps to improve their population.

History and Development

The Marwari Horse was created in India, though their exact origin is yet not known. Genetic researchers opine that they had descended from the Arabian horses crossed with the native Indian ponies and might have some influence from the Mongolian horses as well. However, the Marwari horse share a common genetic structure with their fellow Indian breed, the Kathiawari, while both these horses seem to be carrying the blood of the Arabian breed.

These horses carry some sign of pure blood since centuries, the sign of which is said to be reflected through its ear tips that are pointed inwards to the level that, in some, they even tend to overlap or touch each other.

According to the traditional accounts, the Marwari horses were bred since at least the 12th century. The Marwaris were primarily war horses that were primarily created with the purpose of appearing in the historic battlefields when there was an urgent need of creating such an animal that can act as a hardy mount to the royals and the soldiers. However, other than as war mounts, they proved to be useful for other purposes to the people of the region.

With the end of British rule, the number of these horses became somewhat stable. According to a government survey, around 500 to 600 individuals remained. Though this number was relatively higher than what the scenario had been, a temporary ban under a 1992 biological conservation pact was introduced in order to prevent them from being exported.

Soon after, an association in Chopsani, in Rajasthan's Jodhpur named the Marwari Horse Breeding and Research Institute was established. The institute conducts different educational programs to promote, improve, and maintain these horses and register them with financial aid from the present-day Indian government.

Interesting Facts

- A famous Sanskrit epithet about a unique feature of the Marwari horse is about its neck, Mayura Greeva, which means, having a 'proud, arched neck like a peacock' (interestingly, peacocks are also common in the region).
- A few of the 20th century notable Marwari horses were Nagraj, Alibaba, Alishan, and Adam.
- The Marwari horse gets its name from the historic 'Marwar' region of Rajasthan, in the western-most corner of India.

- Marwari horses were directly associated with the then-prevailing caste system of the country. This means that people, other than the warrior class (Kshatriyas and Rajputs), were not permitted to ride the Marwari horse.
- In the rural regions of Rajasthan, the horse is often trained for 'dancing' on many occasions, marriages and festivals that take place throughout the year.
- Population is 33,698 as per 20th Livestock Census (2019). The percentage (%) share with respect to the total is 9.8, is the highest in India.

Fig. 4: Marwari horse breed

5. Spiti

The Spiti horse is a small-statured equine breed, hailing from the Himachal Pradesh region of the northern part of India. Taking their name from the Spiti river, where they were bred, these horses were widely used as pack animals because of their ability to pursue long journeys while carrying heavy loads on their back quite comfortably.

History

Though no concrete details have been provided regarding the evolution and ancestors of these horses, legend has it that this breed seems to descend from wild horses known as Kiang, inhabiting the Alpine regions of Spiti, Changthang region of Ladakh as well as the pastures of Tibet. However, this perception was changed as the latter belongs to the ass family. In fact, these horses have a close resemblance to the Mongolian and Tibetan breeds that might even be their ancestors. An analysis of genetic diversity has closely linked them to the Zanskari horses, found in the Leh and Ladakh regions of Jammu and Kashmir.

They are mostly found in the Kullu, Spiti, Kinnaur and Lahaul districts of Himachal Pradesh. However, these horses are not well adapted to thriving in very high altitudes. According to certain reports, the population of this breed was said to be around 4000 in the year 2004. In 2007, the Food and Agricultural Organization or FAO mentioned its conservation status to be "not at risk". Lately, the Spiti horses have been marked as "endangered". The breeding of these horses was done in a traditional way, where the villagers used a particular stallion to cover all the mares, with the stallions being changed every year.

Population is 1,130 as per 20th Livestock Census (2019).

Fig. 5: Spiti horse breed

6. Zanskari

The Zanskari horse, often referred to as the Zaniskari, is a distinct breed of small, robust equines that originate from the rugged and lofty landscapes of Ladakh in northern India. This breed, which takes its name from the serene Zanskar region within Kargil district, stands as a testament to the area's natural and cultural richness.

Currently, the Zanskari horse is facing a critical situation with their population numbers dwindling to a mere few hundred. This alarming scenario has prompted the initiation of a focused conservation effort in Padum, located in the Zanskar valley of Kargil district, Ladakh. The primary goal of this program is to safeguard the survival of this unique breed, ensuring that it continues to thrive and remain a vital part of Ladakh's heritage.

History

Tracing the history of the Zanskari horse reveals an intricate interweave of ecological adaptation, cultural significance, and modern challenges. Native to Ladakh in India's high-altitude territories, this breed has experienced both periods of flourishing as well as critical endangerment over its lifetime.

The Flourishing Era (Pre-1977)

Prior to 1977, Zanskari horses were an increasingly common sight across Ladakh's mountainous terrain and estimates pegged their numbers between 15,000 to 20,000, signifying an active and thriving population.

Initial Decline (Post-1977)

As the years passed post-1977, their numbers started dwindling slowly due to various contributing factors. Cross-Breeding Impact: One of the primary contributors to Zanskari horses decline has been their inexact crossbreeding with other breeds, which dilutes their genetic purity.

Habitat Shifts: With the expansion of roads and mechanization in Ladakh, traditional grazing patterns were disturbed, further impacting its population.

FAO Assessment (2007)

Status Designation: In 2007, the Food and Agriculture Organization (FAO) assessed Zanskari horses as "not at risk", an assessment which may have underestimated their vulnerability.

Continued Population Decline (2006): Despite declining numbers, a 2006 study discovered no genetic bottleneck. This suggests a resilient genetic diversity within the breed.

Sharp Decline (2013-2022): By 2013, the population had decreased to approximately 9,700 animals. As time progressed, however, its decline escalated until by 2022 the Domestic Animal Diversity Information System (DAD-IS) reported only 346 individuals left and labelled them "at Risk/ Critical Maintained".

Genetic Studies and Relations

2007 Genetic Analysis: According to research done in 2007, it was discovered that the Zanskari breed shares close genetic ties with Manipuri, Spiti, and Bhutia breeds and distantly with Marwari breed.

2014 Study Findings: In 2014, further research reaffirmed these connections between Zanskari and Spiti breeds in particular.

Conservation Efforts

Breeding Program in Padum: Due to an alarming decrease in their numbers, Jammu and Kashmir's Animal Husbandry Department initiated a conservation and breeding program in Padum, Zanskar in order to safeguard genetic integrity and ensure survival of this unique breed.

Cultural and Ecological Importance

Zanskari Horse's Significance Extends Beyond Figures: The significance of Zanskari horses goes well beyond mere population statistics; this breed is interwoven into Ladakh's cultural fabric, serving as a powerful symbol for its history, traditions, and ecological balance.

Future Prospects and Challenges: Ongoing conservation efforts combined with increased awareness present a ray of hope for the Zanskari horse. Though its future may pose formidable obstacles, the commitment shown towards saving this unique breed speaks to a larger narrative of ecological preservation as well as cultural preservation.

Overall, the Zanskari horse's history serves as an important reminder of how nature and humanity must coexist peacefully and sustainably. Furthermore, it emphasizes the significance of sustainable practices and conservation strategies in maintaining not just breeds but also regions' unique heritage and biodiversity.

Characteristics

The Zanskari horse, native to Ladakh's high alpine landscapes, is revered for its robust yet well-proportioned build; an adaptation perfectly tailored for Ladakh's hypoxic environment.

Stature and Build

Hardy Zanskari horses typically stand from 120 to 140 cm (12-14 hands), boasting compact yet muscular forms due to the rugged terrain they inhabit. These sturdy forms may be seen grazing across different regions around the world.

Measured Proportions: The breed typically displays a thoracic circumference spanning 140 to 150 cm (55 to 60 inches), indicative of its strong and expansive chest. Their body length typically falls within 95-115 cm (38-45 inches), further reinforcing their well-balanced and resilient structure.

Adaptation to High-Altitude Living

Zanskari Horses Are Highly Efficient at Working in Hypoxic Conditions: Zanskari horses have proven themselves remarkably efficient at operating in Ladakh's high altitude environments, where oxygen levels tend to drop drastically. Their physiological traits enable them to complete laborious tasks where other breeds might struggle.

Coat Characteristics

Predominant Grey Hue: For Zanskari horses, grey is often their coat colour of choice, an aesthetic trait as well as natural camouflage within their native terrain.

Diverse Colour Palette: While grey dominates, Zanskaris come in various shades such as bay, brown, black and chestnut; this wide spectrum of colours adds allure and individuality.

The Zanskari horse, with its distinct physical traits and remarkable adaptation to its environment, stands as an embodiment of nature's resilience and versatility.

Not only are these horses physically strong; they also represent life at high altitudes where native species flourish under extreme environmental conditions, further cementing its symbolic role. As such, its presence serves as both practical relevance as well as living evidence of life harmony within Ladakh's mountainous ecosystems.

Temperament and Trainability

Zanskari horses are beloved companions. Easy to train and immensely loyal, their gentle temperament is ideal for riding by novice equestrians alike.

Their intelligence allows them to remember paths and directions more readily, helping them navigate mountainous landscapes safely.

Population is 6,660 as per 20th Livestock Census (2019). The percentage (%) share with respect to the total is 1.9, which is the third highest in India.

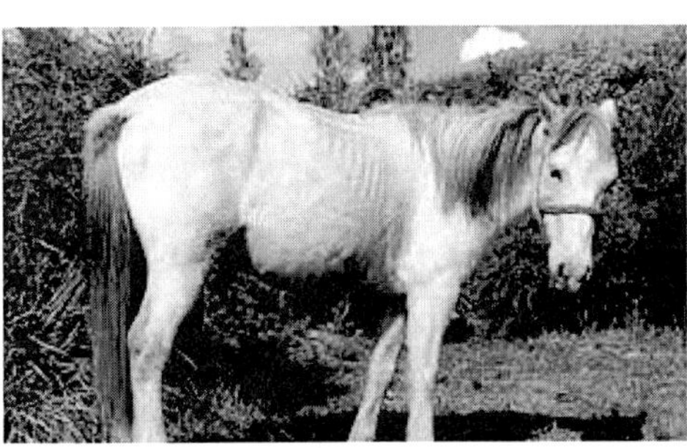

Fig. 6: Zanskari horse breed

7. Kachchhi-Sindhi

- Kachchhi-Sindhi horses are famous for their 'Rewal chal' (a unique style of running). These horses dominantly exist in the western-northern border of India adjoining Pakistan.

- The Kachchhi-Sindhi breed is a complete desert horse, scientists say. Its broader hoofs make it easy to tread through desert sand while the covered nostrils and strong stamina allow it to perform in tough conditions. Aficionados have often called this breed 'Mustangs of India', though it is not a feral or mixed breed.
- The breeding tract of these horses is **Surat, Navsari, Kachchh district of Gujarat and Jaisalmer-Barmer districts of Rajasthan in India**.
- Most familiar colours in the Kachchhi-Sindhi horses are bay and chestnut.
- Roman nose, ears curved at tips but not touching each other, short back, short pastern length, broader hoof for better grip and docile temperament are major characteristics of these horses.
- Kachchhi-Sindhi breed Ears are curved at tips but not touching each other whereas horses of both Marwari and Kathiawari Indian breeds can rotate their ear tips by 180°, leading the tips to meet together at the open end.
- It is famous for its 'Rewal chal' royal style of running.
- **Rewal in itself is a different version of trot and is faster and comfortable for the rider.** As per the belief of breeders, it runs rewal faster than the Marwari horses.
- The horses are docile in temperament.
- Fourteen biometric indices were recorded for phenotypic characterization of each breed viz., height at wither (HW), body length (BL), heart girth (HG), neck length (NL), face length (FL), face width (FW), pole, ear length (EL), ear width (EW), foreleg length (FLL), canon, pastern, hoof length (HoL) and hoof width (HoW).
- On average, these horses stand 148 cm height, have a body length of ~140 cm, a heart girth of ~165 cm, an ear length of ~15 cm and a face length of ~61 cm.
- Horse keepers sustain horses in intensive as well as an extensive system of rearing.
- It has also been ascertained from various sources that horse number is declining rapidly, however, breed population statistics are not available.
- There is, therefore, an urgent need to conserve this breed.

- Proper managemental practices and conservation efforts will pave the way for the multiplication of this valuable equine genetic resource of India.
- Population is 3,943 as per 20th Livestock Census (2019). The percentage (%) share with respect to the total is 1.2, which is the fourth highest in India.

Fig. 7: Kachchhi-Sindhi horse breed

8. Bhimthadi

- It is a newly registered eighth breed of horse by NBAGR.
- **Bhimthadi horse** is distributed in **Pune, Solapur, Satara and Ahmednagar district of Maharashtra**. The average height of a stallion is about 130 cm and of mare is 128 cm.
- The predominant coat colour is liver chestnut.
- They are used for transportation of household material during migration of pastoral communities in the region.
- Average adult body weight of a stallion is 267 kg.

Fig. 8: Bhimthadi horse breed

9

Donkey Breeds of India

Donkeys can also be termed as ass (***Equus africanus asinus***) and is a domesticated member of the horse family, ***Equidae*** with chromosome number 62. The Wild ass of Africa, ***E. africanus*** is the wild ancestor of the donkey. For approximately 5000-7000 years, the donkey has been used as a working animal.

An adult male donkey is a jack or jackass, an adult female is a jenny or jennet, and an immature donkey of either sex is a foal. Jacks are often mated with female horses (mares) to produce **mules**, the less common hybrid of a male horse (stallion) and jenny is called **hinny**. NBAGR-ICAR has registered three breeds of donkey till date.

1. Kachchhi

- These donkeys are found in Kachchh district of Gujarat.
- Coat colour is mainly grey (dorsal surface grey and ventral surface white) followed by white, brown and black.
- Forehead is convex. Nasal bone is straight.
- Height at wither ranges from 77 to 115 cm.
- Docile in temperament.
- Only donkeys are used for agricultural purposes like inter cultivation for weed removing.
- Also utilized for transportation as donkey cart, as pack animal during pastoralist migration, etc.
- Can carry approximately 80-100 kg on back and can pull 200-300 kg on cart. Population size is approximately 1700.

Fig. 1: Kachchhi breed of donkey

2. Spiti

Breeding tract: Kinnaur, Lahaul and Spiti (Himachal Pradesh)

Comment on breeding tract: The breeding tract of Spiti donkeys is a high altitude, mountainous area situated between 3000 to 4200 meters above mean sea level and comprises the Spiti subdivision of Lahaul-Spiti and Pooh subdivision of Kinnaur district. The vegetation is cold arid alpine type.

Main use

- Transportation
- The Spiti donkeys are very strong and sure-footed animals with incredible stamina.

Origin: the breed is known since time immemorial and is named after the name of its native habitat.

Adaptability to environment

They can survive well in scarcity of feed and fodder during harsh winter months when this area is completely snow bound. Spiti donkeys are able to carry up to 100 kg of load in high altitude mountainous areas with thin atmosphere having lesser oxygen.

Colour

Mainly brown, also brown-black and black. Major white markings mostly around the muzzle and sometimes around eyes.

Characteristics

- The face is also covered with long hairs like the rest of the body.
- The head is comparatively broader and shorter.
- Tail extends up to the hocks.

- Tail switch is not distinguishable due to the presence of long hair on the rest of the tail.
- The Spiti donkeys are comparatively smaller in size having an average height of less than 90 cm. They are strongly built with compact bodies with strong and straight backs. They have a thick coat of hair all over their body.

Fig. 2: Spiti breed of donkey

3. Halari

- **Native to Saurashtra region of Gujarat**.
- These donkeys are white in colour. Muzzle and hooves are black. Forehead is mostly convex.
- Halari donkeys have a strong built and large size with an average height at wither of 108 cm in males and 107 cm in females, and average body length of 117 cm in males and 115 cm in females.
- These donkeys are very docile in temperament, and are used as pack animals during pastoralist migration and for transportation as donkey carts.
- Halary Donkeys can walk approximately 30-40 km in a day during migration.

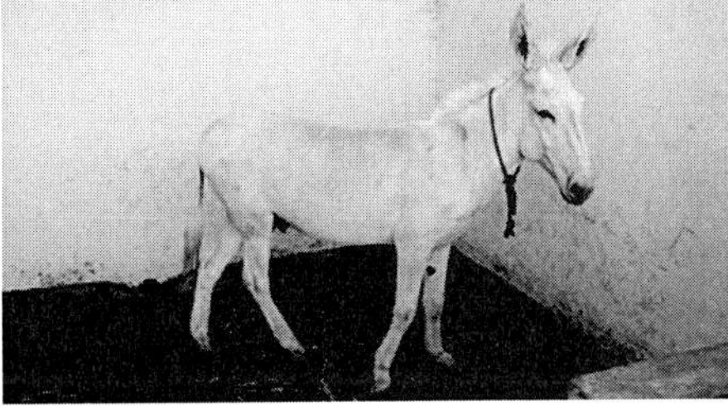

Fig. 3: Halari breed of donkey

10

Dog Breeds of India

The full name of an organism technically has eight terms. For the dog it is: Eukarya, Animalia, Chordata, Mammalia, Carnivora, Canidae, Canis, and lupus. Notice that each name is capitalized except for species, and the genus and species names are italicized. Scientists generally refer to an organism only by its genus and species, which is its two-word scientific name, or binomial nomenclature. Therefore, the scientific name of the dog is ***Canis lupus***. The name at each level is also a taxonomy Carnivora is the name of the taxon at the order level; *Canidae* is the taxon at the family level, and so forth. Organisms also have a common name that people typically use, in this case, dog. Note that the dog is additionally a **subspecies: the "familiaris"** in ***Canis lupus familiaris***. Subspecies are members of the same species that are capable of mating and reproducing viable offspring, but they are separate subspecies due to geographic or behavioural isolation or other factors. Dogs, which are under the species *Canis lupus familiaris*, are known to have a total of 78 chromosomes (2n). The diploid genes can be classified into 38 pairs with two sex chromosomes.

The dog (*Canis familiaris* or *Canis lupus familiaris*) is a domesticated **descendant of the wolf**. Also called the domestic dog, it is **derived from extinct grey** wolves, and the grey wolf is the dog's closest living relative. The dog was the first companion species to be domesticated by humans. Hunter-gatherers did this over 15,000 years ago, which was before the development of agriculture. As dogs have long association with humans, dogs have expanded to a large number of domestic individuals and gained the ability to thrive on a starch-rich diet that would be inadequate for other canines. NBAGR-ICAR has registered three breeds of dog till date.

1. Rajapalayam

- Rajapalayam dog is mainly distributed in **Virudhunagar, Tirunelveli and Madurai districts of Tamil Nadu**.

- These dogs are medium in size with a compact body. Coat colour is white. Skin, nostrils and eyelids are pink. Eyes are golden.
- Height at wither ranges from 55 to 72 cm in males and 38-70 in females.
- Adult body weight ranges 14 to 32 kg. Rajapalayam dogs are utilized for guarding farms and farm houses.
- These possess the attributes of high obedience and easy trainability. **Barking is medium pitched. Estimated population is 3000-4000.**

Fig. 1: Rajapalayam dog breed

2. Chippiparai

- Chippiparai dogs are mainly found in **Thoothukudi, Tirunelveli, Virudhunagar and Madurai districts of Tamil Nadu**.
- Estimated population is about 6000. Chippiparai dogs are also called Kanni (virgin) or Vettainaai (hunting dogs).
- These are medium in size. Coat colour varies from fawn to dark brown, brownish black and black. Black dogs have white markings on both sides above the eyes, or black circle around the eyes.
- Eyes are golden and oval. Ears are medium in size and drooping or semi drooping.
- Height at wither ranges from 60 to 76 cm in males and 54 to 70 cm in females. **Adult body weight ranges from 13.6 to 32.5 kg.**

- **The utility of this dog is mainly for guarding and hunting, but also kept as a hobby and pride by the owners.**
- **These dogs are high in obedience and easy to train.**

Fig. 2: Chippiparai dog breed

3. Mudhol Hound

- Mudhol Hound dogs are also known as **Pissouri Hound or Lahori Hound. These dogs are mainly distributed in Bagalkot and Vijayapur districts of Karnataka**.
- Estimated population is about 1500.
- They are strongly built, having high stamina and endurance. Body is symmetrical with an elegant and lean look.
- Coat colours include mainly white, brown, patchy, brindle, black, fawn along with spotted.
- Head is proportionately small. Skull is long and narrow.
- Eyes are dark brown or hazel coloured, oval and obliquely placed. Ears are medium, thin, triangular and set high. Chest is long and deep. Abdomen is well tucked.
- The tail is long, tapering, slightly curved. Height at wither ranges from 73 to 80 cm in males, and 61 to 74 cm in females.
- **Adult body weight ranges from 21 to 40 kg. Gait of Mudhol Hound, with an aerodynamic body gives an effortless stride, giving a flying appearance.**

- **Mudhol Hound dogs are used as guarding and shepherding. These dogs are also high in obedience and easier to train.**

Fig. 3: Mudhol Hound dog

Foreign breed of dogs

1. German Shepherd

- Worldwide interest in the breed began rising in the early 1900s and the German shepherd was recognized by the American Kennel Club (AKC) in 1908.
- During World Wars I and II, the word "German" was dropped and the breed was referred to as the **shepherd dog or the Alsatian**, a name that is still often used in Europe and India.
- **Characteristics of breed:**
- Group: herding
- Height: 22 to 26 inches
- Weight: 60 to 100 pounds
- Coat: coarse, medium-length double coat
- Coat colour: black and tan, black and cream, black and red, black and silver, solid black, gray, sable. Note that blue, liver or white are unfavourable based on breed standards.

Fig. 4: German Shepherd dog breed

- Temperament: intelligent, courageous, alert, bold, loyal, protective
- Origin: Germany

2. Golden Retriever

- The Golden Retriever is one of the most popular dog breeds in the United States or around the World.
- The breed's friendly attitude makes them great family pets, and their intelligence makes them highly capable working dogs.
- **Origin**: Scotland
- Size: 55-75 pounds (25-34 kg) for females and 65-75 pounds (29-34 kg) for males.
- **Lifespan**: 10-12 years
- **Coat**: Dense, water-repellent double coat that is usually golden in colour. They can have different shades of gold, from pale to dark.
- **Temperament**: Golden Retrievers are friendly, intelligent, and devoted dogs. They are known for their gentle and patient nature, making them excellent family pets and service dogs.

Fig. 5: Golden Retriever dog breed

3. Labrador Retriever (Lab)

- The Labrador retriever is a medium to large dog breed with a short coat and sturdy physique that's a member of the sporting group and originated in Newfoundland and the U.K.

Fig. 6: Labrador Retriever dog breed

- Labs are known for their intelligence, fine character, and good temperament. Although they were bred to be hunting dogs, they are also excellent companions.
- Group: Sporting
- Height: 22.5 to 24.5 inches (males), 21.5 to 23.5 inches (females)
- Weight: 65 to 80 pounds (males), 55 to 70 pounds (females)
- Coat: short, dense double coat

- Coat colour: Various shades of gold, black, chocolate, or yellow
- Life span: 10 to 12 years
- Temperament: Friendly, active, companionable
- Hypoallergenic: No
- Origin: Newfoundland/United Kingdom

4. Pug

- The Pug, a charming and distinctive breed of dog, boasts a rich history that stretches back centuries.
- Originating in China during the Shang dynasty (around 1600–1046 BCE), Chinese royalty cherished Pugs for their endearing personality and elegant appearance.
- With a compact yet robust body, a short, wrinkled muzzle, and a curled tail, Pugs quickly became favourites of European nobility after they were introduced to the continent in the 16th century.
- Origin: China
- Size: Small
- Breed group: Toy group
- Lifespan: 12-15 years

Fig. 7: Pug dog breed

- Coat: Short and smooth
- Temperament: Pugs possess an affectionate, charming, and playful temperament. They often act as "clowns" and establish friendly relationships with both people and other pets. They build strong bonds with their owners.

5. Spitz

- A spitz (the name derives from the German word spitz, (German: meaning "pointed", in reference to the pointed muzzle.
- The tail often curls over the dog's back or droops.
- While all of the breeds resemble primitive dogs, smaller breeds resemble foxes, while larger breeds resemble jackals.

Fig. 8: Spitz dog breed

- Group: Non-sporting (AKC)
- Height: 17 to 18 inches
- Weight: 35 to 45 pounds
- Personality/temperament: Friendly, affectionate, people-loving
- Energy level: Medium
- Coat and colour: Famous "spectacle" markings around the eyes; very thick double topcoat, woolly, undercoat, and long outer coat; comes in cream, black, and gray colour variations
- Life span: 12 to 15 years

6. Rottweiler (Rottie)

- The Rottweiler, or Rottie, is a large and rugged dog with a hard-working and confident demeanour.
- It is known for its muscular body, thick hindquarters, and easy trot.
- Descended from the mastiffs of the Roman legions, the Rottie can be a gentle playmate and companion, despite its reputation for being dangerous.
- The breed is a bit aloof but also very intelligent, protective, and loyal to its family.

Fig. 9: Rottweiler dog breed

- Group: Working
- Height: 22 to 27 inches at the shoulder
- Weight: 80 to 130 pounds
- Coat: Short double coat
- Coat colour: Black with tan, rust, or mahogany
- Life span: 8 to 10 years
- Temperament: Steady, alert, self-assured, fearless, devoted, confident, good-natured, obedient

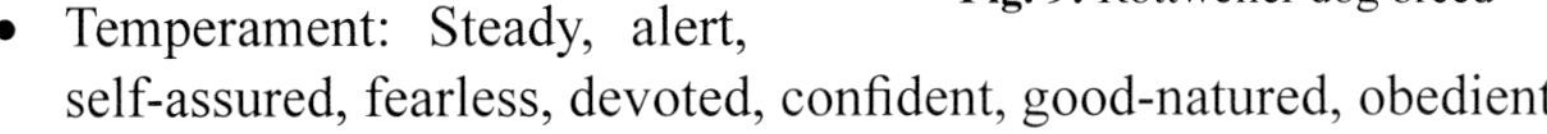

- Hypoallergenic: No
- Origin: Germany

7. Pomeranian

- It is a tiny dog, a toy breed weighing just several pounds, with a long coat and distinctive neck ruff that harken back to their spitz heritage.
- Poms are typically friendly but can be bossy and barky, so they're not the dog for everyone.
- Group: Toy (American Kennel club)
- Height: 6 to 7 inches
- Weight: 3 to 7 pounds
- Coat: Long double coat
- Coat colour: Comes in many colours and combinations, though the most common are red, orange, cream, sable, black, brown, and blue
- Life span: 12 to 16 years
- Temperament: Bold, alert, lively, affectionate
- Hypoallergenic: No
- Origin: Germany and Poland

Fig. 10: Pomeranian dog breed

8. Dalmatian

Fig. 11: Dalmatian dog breed

- It is a sleek and medium-sized short haired dog that is affectionate, very athletic, highly intelligent, and energetic.
- This breed is extremely driven, loyal, and well-suited for work or companionship.
- The Dalmatian is known for its distinct spots, slim carriage, and tail arched upward.
- It also has a rich history as a firehouse dog and, of course, its acclaim from the Disney cartoon movie.
- Group: Non-sporting
- Height: 19 to 24 inches tall
- Weight: 45 to 65 pounds
- Coat: Short
- Coat colour: White with black or liver spots
- Life span: 11 to 13 years
- Temperament: Outgoing, friendly, intelligent, active, energetic, sensitive, playful
- Hypoallergenic: No
- Origin: Possibly Dalmatia/Croatia or England

9. Irish Setter

- Origin: Irish Setters are a gundog breed that originated in Ireland. Originally bred to hunt birds, some Irish Setters continue hunting them today.
- Size: Irish Setters are large dogs, typically standing 25-27 inches tall at the shoulder and weighing 60-70 pounds. They have long, silky coats and feathered ears.
- Lifespan: Irish Setters have a lifespan of 12-15 years. Ultimately, they are generally healthy dogs, but some common health problems include elbow and hip dysplasia, eye problems, and cancer.

- Coat: The Irish Setter's coat is long, silky, and flowing. Though they can be any colour, the most common colours are red, mahogany, and chestnut. Brush the coat daily to prevent matting.

Fig. 12: Irish Setter dog breed

- Temperament: Moreover, Irish Setters are friendly, outgoing, and intelligent. They are playful and affectionate dogs that love to be around people.

10. Doberman Pinscher (Dobie)

- The Doberman pinscher is a medium-large, deep-chested dog breed with a sleek and sturdy appearance.
- Dobermans (also called "Dobes" or "Dobies") are fearless, loyal, and highly intelligent. These traits make them ideal police, war, and guard dogs, but they are also outstanding companions.
- Group: Working
- Height: 24 to 28 inches
- Weight: 65 to 100 pounds
- Coat: Short and smooth
- Coat colour: Black, red, blue, or fawn with rust markings (sometimes small patches of white are seen)
- Life span: 10 to 12 years
- Temperament: Intelligent, loyal, alert, energetic, attentive
- Hypoallergenic: No
- Origin: Germany

Fig. 13: Doberman Pinscher dog breed

11. Dachshund (Doxie)

- The dachshund is an energetic, lovable small dog breed from Germany with an endearing personality and is known for its varied coat texture and colour, short legs, floppy ears, and big chest.

Fig. 14: Dachshund dog breed

- It's affectionately called a Doxie, wiener dog, hotdog, or sausage dog, and this cute pint-size breed definitely leaves a lasting impression.
- Group: Hound
- Height: 8 to 9 inches (standard); 5 to 6 inches (miniature)
- Weight: 16 to 32 pounds (standard); up to 11 pounds (miniature)
- Coat: Varieties include smooth (short haired), longhaired, and wire-haired.
- Coat colour: Colours include (but not limited to) black, tan, fawn, beige, blue, chocolate, and red with various markings such as dapple, piebald, brindle, and sable.
- Life span: 12 to 16 years
- Temperament: Clever, playful, stubborn, devoted, lively, independent, courageous
- Hypoallergenic: No
- Origin: Germany

12. Siberian Husky

- Best For: Family pet and security/guard dog
- Colour Options: Mix of white and grey
- Characteristics: Intelligent, friendly, gentle, alert
- Life Span: Up to 15 years
- The popularity of the Siberian Husky breed of dogs in India has tremendously increased over recent years.

- They are classic Northern dogs with natural coats to protect them from winter, thus, making them ideal for cold temperatures.
- The Siberian Husky was bred in Siberia, primarily for being a companion, guard, and sled dog. When they were used to transport antitoxins to Nome, Alaska, to fight the diphtheria epidemic, their popularity grew.

Fig. 15: Siberian Husky dog breed

- The Siberian Husky breed of dogs has thick manes with a medium-length double coat that can withstand colder temperatures.
- They can be found in grey to pure white shades, and sometimes even brown and white. They have erect ears, straight necks, and a level top line.

13. Boxer

- Best For: Family pet and security/guard dog
- Colour Options: Brown, black, and off-white
- Characteristics: Devoted, loyal, bright, fearless
- Life Span: Up to 12 years
- Boxer dogs are known for their energy-filled nature.
- This dog's inherent patience and protective nature have earned it a reputation as a great dog with children. They can be found in colours like brindle, fawn, and white.
- Their short and shiny coat requires very little grooming. They can grow up to a height of 25 inches.

Fig. 16: Boxer dog breed

- Boxer dogs are also known for their strength and agility. They make good watchdogs for families.
- While they show patience and playfulness towards children, they respond aggressively to anything that threatens their loved ones.

14. Beagle

- Best for: Family pet
- Colour options: Mix of brown, black, and white
- Characteristics: Determined, excitable, amiable, gentle
- Life span: Up to 15 years
- Beagles are known for being gentle, especially with children. They are fun-loving and energetic, too, which also makes them great family dogs.

Fig. 17: Beagle dog breed

- Beagles are extremely friendly and intelligent dogs. They're the most adorable and one of the preferred pet dog breeds worldwide.
- This breed of dog usually grows up to a maximum height of 15 inches. They are small to medium-sized dogs, and often, are used as sniffer dogs in airports because of their strong sense of smell.
- This breed of dogs is available in a mixture of three colours brown, black, and white.

15. Greyhound

- A famous racing breed, the intelligent Greyhound is highly energetic. However, they are pretty sensitive and need a soft hand in training. If training is too harsh, they can become timid and fearful.
- The Greyhound is a gentle and intelligent breed whose combination of long, powerful legs, deep chest, flexible spine, and slim build allows it to reach average race speeds exceeding 64 km per h (40 mph).
- The dog's narrow head, long neck, and deep chest are also recognizable aspects of this breed. These physical characteristics make the greyhound the fastest breed in the world.

- Group: Hound
- Height: Male 28 to 30 inches, Female 27 to 28 inches
- Weight: 60 to 80 pounds
- Coat: Short and smooth
- Coat colour: Black, blue, fawn, red, white, and various shades of brindle, or a combination of any of these colours
- Life span: 10 to 13 years
- Temperament: Even-tempered, intelligent, affectionate, athletic, quiet, gentle
- Hypoallergenic: No
- Origin: Egypt

Fig. 18: Greyhound dog breed

16. Chihuahua

It is a Mexican breed of toy dog. It is named for the Mexican state of Chihuahua and is the **smallest dog breed in the world**, and also has the longest life expectancy of up to 20 years plus.

Breed Overview

Group: Toy

Height: 5 to 8 inches

Weight: Not exceeding 6 pounds

Coat: Smooth and short or longhaired

Coat color: Black, black and tan, blue and tan, chocolate, chocolate and tan, cream, fawn, fawn and white, red

Life span: 14 to 16 years

Temperament: Loyal, alert, lively, attentive, bright, companionable

Origin: Mexico

They claim distinct characteristics, like wide eyes and ears that are usually erect and very large in relation to their small head and body. The Chihuahua has a exceptional personality and can be an loving and loyal companion dog as long as they don't feel threatened. This breed was first registered by the American Kennel Club (AKC) in 1904, the Chihuahua is one of the oldest breeds on the American continent and one of the smallest breeds in the world.

Fig. 19: Male and female Chihuahua dog breed

11

Yak Breeds of India

The yak is a bovine species that provides livelihoods for people in high altitude conditions of extreme harshness and deprivation. Yaks live predominantly on the "**roof of the world**", as the **Qinghai-Tibetan Plateau** is often called, and provide milk, meat, hair and down fibre, hides, draught power and dung (principally used as fuel). The yak has physical and physiological characteristics that allow it to thrive at high altitudes (low oxygen) and in extreme cold (at temperatures as low as -40 °C) and to survive feed shortages in the winter (FAO).

The milk yield of the yak cow is often no more than the amount suckled by the calf and is not comparable to the milk yield of dairy cattle. However, although milk offtake for human consumption may be at the expense of the calf, yak milk is important for households. In economic terms, milk is often the most important yak product. Yak milk is generally produced by small-scale farmers in traditional systems where management is highly influenced by climate and seasons. Lactation is seasonal and yak cows produce between 150 and 500 litres; yields vary by breed and location. Lactation can generally continue into a second year without another calving. During the winter, yak cows do not go dry and continue to produce a small amount of milk, with milk yields as low as 2 to 4 litres/month. During the second year of lactation, milk yields are between half and two-thirds of those in the first year. Currently, there are no specialized dairy yak breeds (FAO).

Yak Breeds of India

Indigenous domestic yak (*Poephagus gruninens*) of Arunachal Pradesh has probably originated from its wild counterpart (*Poephagus mutus*) long back over the centuries.

- In India, the yaks are reared by the people residing between 3,000 and 6,000 meters above mean sea level in the states of **Jammu-Kashmir, Himachal Pradesh, Uttrakhand, Sikkim, Arunachal Pradesh and West Bengal.**

- Yaks thrive well in these extremely cold regions with hypoxic conditions providing milk, meat, hide, fibre, manure and draught power to the local population.
- They are able to withstand the fodder scarcity endemic to these regions in the winter months. The total yak population in our country is 0.76 lakhs.
- Livestock Census (2019) reported approximately 58,000 Yaks in the country, marking a **25% decline compared to the 2012 census.**
- Jammu and Kashmir has the maximum population of yak (0.54 lakh), possessing about 71% of the total Yak population of the country, followed by Arunachal Pradesh (18.34%) and Sikkim (5.26%).
- The milk yield per lactation in indigenous Yaks is 250–500 kg in a lactation period of 260–300 days.
- The birth weight of indigenous yaks in males and females is 14.20±0.33 and 13.20±0.34 kg, which increases to 201.23±4.79 and 178.13±3.26 kg, respectively, at two years of age.
- Mainly, four types of Indian yaks have been described namely **Arunachali, Ladakhi, Sikkimi and Himachali.**
- Out of these, **Arunachali yaks are the first and only recognized breed of Indian Yaks**.
- This review summarizes the population trends, attributes of indigenous Yaks in general as well of specific types of indigenous yaks available in India.

The diploid number of metaphase chromosomes in male as well as in female yak to be 2n= 60, out of which 58 are autosomes and 2 are sex chromosomes. From the centromeric position, karyotyping studies revealed that all the 29 pairs of autosomes were acrocentric and the sex chromosomes, X and Y were submetacentric and metacentric, respectively.

1. **Arunachali yak** breed is reared by tribal yak pastoralists known as Brokpas who migrate along with their yaks to higher reaches (at an altitude of 10,000 ft and higher) during summers and descent to mid-altitude mountainous regions during winters. It has a population size of 14,061 as per 20th livestock census.

Fig. 1: Arunachali yak breed

12

Mithun (Gayal) Breeds of India

Some have described Mithun as **domesticated Gaur** others as a hybrid descent from the **crossing of gaur bulls and common cows**. Considered to have originated along the Assam and Myanmar border.

- Mithun, also known as 'Cattle of Mountain" is an important bovine species of north-eastern hill region of India and also of China, Myanmar, Bhutan and Bangladesh. This magnificent massive bovine is presently reared under free-range condition in the hill forests at an altitude of 1000 to 3000 m above mean sea level.
- Mithun plays an important role in the socio-economic and cultural life of the local tribal population.
- Presently, this animal is mainly reared for meat, which is considered to be more tender and superior over the meat of any other species.
- Mithun milk, though produced less in quantity, is of high quality and can be used for preparation of various milk products.
- Leather obtained from this species has been found to be superior to cattle.

The karyotype of the mithun (*Bos frontalis*) consists of 58 chromosomes ($2n = 58$). This includes 54 acrocentric autosomes and 4 large submetacentric chromosomes. This karyotype distinguishes mithun from cattle, which have 60 chromosomes ($2n = 60$). Complete cytogenetic analysis including karyotyping and different chromosomal bandings (C-banding and R-Banding) carried out revealed that the normal diploid number of Mithun was 58XY and 58 XX for males and females, respectively. In order to find out the karyotypic evolution of Mithun, the FISH technique was used on the metaphase chromosome of Mithun as well as the wild ancestral species, Gaur. Besides, several economically important genes including kappa casein, leptin, and growth hormones were also characterized. In the recent past, this section also carried out the microsatellite-based characterization of different Mithun populations and muscle transcriptome analysis.

Origin and Distribution of Mithun

Mithun is believed to have originated more than 8000 years ago and is considered to be

Descended from **wild Indian gaur** (Simoons, 1984; Mondal and Pal, 1999). Mithuns are found over a large area of Southeast Asia. Beside meat, Mithuns are reared for sacrificial purposes and/or for barter trade. Their natural habitat is the forests of highlands. In some folk tales, Mithun has been said to be the descendent of the Sun. Different interesting and divergent legends are available on the origin of Mithun among different tribes. Even today, Mithun is used as a holy sacrificial animal to appease the Gods by the tribesman. Mithun, a unique bovine species, has a limited geographical distribution. It is mainly found in the tropical rain forests of North Eastern hilly states of Arunachal Pradesh, Nagaland, Manipur and Mizoram of India

Table 1: Recent trends in Mithun population in India

State	**1997**	**2003**	**2007**
Arunachal	124,194	184,343	218,931
Nagaland	33,445	40,452	33,385
Manipur	16,660	19,737	10,024
Mizoram	2,594	1,783	1,939
Total	**176,893**	**246,315**	**264,279**

Population growth

- The four states in Northeast India, Arunachal Pradesh, Nagaland, Manipur, and Mizoram, have an estimated 3.9 lakh Mithuns.
- **This represents a 30% increase over the 2012 census.**
- **Arunachal Pradesh has the highest increase, with 3,50,154 Mithuns in 2019 compared to 2,49,000 in 2012.**
- However, Nagaland and Manipur have recorded negative population growth of (minus) 33.69% and 10.58%, respectively.
- In Nagaland, the Mithun population dipped to 23,123 in 2019 from 34,871 in 2012.
- In Manipur, the population dropped to 9,059 in 2019 from 10,131 in 2012.
- **On a positive note, there was a 20.38% growth in the Mithun population in Mizoram, reaching 3,957 in 2019 from 3,287 in 2012.**

Strains of Mithun

As described by Verma (1996), two distinct types of Mithuns are available in India and they were named after the name of the State, where they belong (Nagami and Arunachali). These two distinctive types have also been reported by Arora (1998). However, Bhusan *et al.* (2000) have identified four distinct strains of Mithun and named them as **Arunachalee, Mizorami, Nagami and Manipuri strains.**

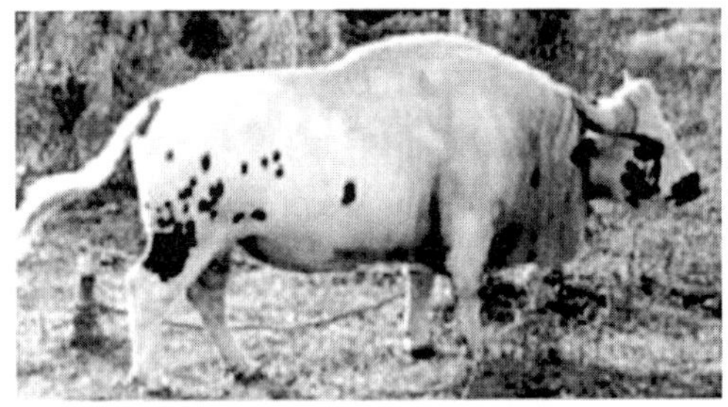

Fig. 1: Arunachalee breed

Fig. 2: Manipuri breed

Fig. 3: Nagami breed

Fig. 4: Mizorami breed

Conservation of Mithun

Keeping in view the dwindling population of Mithun over the years, it is of great priority for the Mithun inhabited states to conserve and propagate quality Mithun germplasm at a faster rate to stabilize its population. There are three ways for the conservation of Mithun genetic resources:

i) through cryopreservation of genetic material like living ova, embryos or semen;

ii) preservation of genetic information as DNA; and

iii) conservation of live population (in situ conservation).

Ex situ versus *in situ* methods of conservation for Mithun

Ex situ preservation involves the conservation of Mithun in a situation removed from their normal habitat. It is used to refer to the collection and freezing in liquid nitrogen of animal genetic resources in the form of living **semen, ova or embryos**. It may also be the preservation **of DNA segments in frozen blood or other tissues**. Finally, it may refer to **captive breeding** or other situations far removed from their indigenous environment.

***In situ* conservation** is the maintenance of live populations of animals in their adaptive environment or as close to it as is practically possible. For domestic species the conservation of live animals is normally taken to be synonymous with in situ conservation.

Mithun husbandry in the North Eastern hill region of India is an important component of the livestock production system. Scientific rearing of this species will not only support the need of protein but also help to generate extra income to the poor Mithun farmers for their livelihood. The need of the hour is, therefore, to popularize scientific farming in the states where Mithun rearing is an age-old practice. The recent success in the field of artificial insemination, oestrus synchronization coupled with timed AI and embryo transfer technology will definitely help to go a long way to achieve the target of propagating quality germplasm in the farmers' field.

13

Endangered Breeds of India

Using the FAO criteria, Bargur and Siri cattle breeds and Attapadi goat breeds were categorized under **vulnerable status** of risk; Krishna Valley, Mewati, Pullikulam and Punganur cattle breeds, Chilika and Toda buffalo breeds and Karnah and Poonchi sheep breeds under **endangered status of under risk** and Vechur cattle under **critical status of risk**. All these breeds need immediate steps for rescue by using suitable conservation methodology. Breed-wise census should be continued in future so that trends in breed populations may also be estimated for fine tuning of risk assessment of breeds.

Rate of inbreeding in Indian livestock breeds: The rate of inbreeding estimated for different breeds of cattle was well below 0.5% (needed for a breed 'not at risk') except for Mewati (1.193%), Pullikulam (0.72%) and Vechur (17.893%). The rate of inbreeding in buffalo breeds was also well below 0.5% except in Chilika buffalo, where it was 0.731%. The rate of inbreeding of these breeds may be reduced by increasing the number of breeding males and in Vechur also the breeding females. The rate of inbreeding in all the breeds of sheep (Indian or exotic), goat and pig were estimated to be less than 0.5%. In these species, a good balance of breeding males and females is always expected.

- The number of breeding females of a breed is the main determining factor to decide the endangered status.
- The estimates of minimum population size for different species to declare the endangered status have been given by different workers.
- However, it depends on species, local circumstances like breed management system, rate of crossbreeding.
- Rate of decline and utility of breed.
- The mortality rate.
- Growth rate, sexual maturity, fertility rate in different species are different and hence the Endangered status differs for different species.

- According to FAO, the survival risk of the breed should be studied when breed population size reaches 5000 breedable females and total population of 10,000.

Risk Status Classification of Breeds

As per FAO (2013), the risk status classification for Bovidae family (low reproductive capacity species) has been given. It includes seven categories namely.

1. **Extinct:** When there is no breeding male or female remaining and any cryo-conserved genetic material that may be available is insufficient for breed reconstitution; Cryo-conserved only, when there is no living male or female animals remaining but for which there is sufficient cryopreserved material to allow for reconstitution of the breed.
2. **Critical:** A breed is categorized as critical, if the total number of breeding females is less than or equal to 300 or the overall population size is less than or equal to 240 and the population trend is increasing and the proportion of females being bred to males of the same breed is greater than 80% or the overall population size is less than or equal to 360 and the population trend is stable or decreasing or the total number of breeding males is less than or equal to five (i.e. Δ F is 3% or greater). If the population trend is unknown, then it is assumed to be stable. This may be called as "**Critical-maintained**", if active conservation programmes are in place or populations are maintained by commercial companies or research institutions.
3. **Endangered:** A breed is categorized as endangered, if the total number of breeding females is greater than 300 and less than 3,000 or the overall population size is greater than 240 and less than 2,400 and the population trend is increasing and the proportion of females being bred to males of the same breed is greater than 80% or the overall population size is greater than 360 and less than or equal to 3,600 and the population trend is stable or decreasing or the total number of breeding males is greater than five but less than or equal to 20 (i.e. Δ F is between 1 to 3%). If the population trend is unknown, then it is assumed to be stable. This may be called as "**Endangered-maintained**", if active conservation programmes are in place or populations are maintained by commercial companies or research institutions.
4. **Vulnerable:** A breed is categorized as vulnerable, if the total number of breeding females is between 3,000 to 6,000 or the overall population size

is greater than 2,400 and less than or equal to 4,800 and the population trend is increasing and the proportion of females being bred to males of the same breed is greater than 80% or the overall population size is greater than 3,600 and less than or equal to 7,200 and the population trend is stable or decreasing or the total number of breeding males is between 20 to 35 (i.e. ΔF is between 0.5 to 1%). If the population trend is unknown, then it is assumed to be stable;

5. **Not at risk:** If the population status is known and the breed does not fall in any of above categories; Unknown, if the population data is not available for the breed. It is important to note that pig is considered as high reproductive capacity species, therefore, breedable female population and overall population size of the breed in the above said criteria will be one third of the figures given for low reproductive category species (cattle, buffalo, sheep and goat) and all other criteria will remain the same.

Table 1: The population size in thousands for consideration of endangered status of a breed under Indian conditions for different species have been suggested as under by NBAGR

Species/Status	**Greater than 10000 Normal**	**10000-5000 Insecure**	**5000-1000 Vulnerable**	**1000-100 Endangered**	**Less than 100 Critical**
Cattle	25	15-20	5-15	2-5	Less than 2
Buffalo	30	20-30	10-20	5-10	Less than 5
Sheep and Goat	50	30-50	15-30	8-18	Less than 8
Camels and Horse	20	15-20	5-15	2-5	Less than 2
Pigs	10	5-10	1-5	0.5-1.0	Less than 0.5

Table 2: Breed watch list (NBAGR) 2022

Cattle	**Buffalo**	**Sheep**	**Goat**	**Camels**	**Horse**	**Pigs**
Endangered	**Vulnerable**	**Critical**	**Critical**	**Critical**	**Endangered**	**Endangered**
1. Belahi	1.Chilika	1.Tibetan	1. Teressa	1.Malvi	1.Bhutia	1.Agonda Goan
2. Khariar	2.Toda			2.Mewari	2.Kachchhi-Sindhi	2.Tenyi Vo
3. Krishna Valley				3.Mewati	3.Manipuri	Not at risk
4. Pulikulam	**Not at risk**	**Endangered**	**Endangered**	**Endangered**	4.Spiti	1.Doom
	Rest all breeds	1.Karnah	1.Chegu	1.Jalori	5.Zanskari	2.Ghungroo
Vulnerable		2.Katchaikatty black	2.Sumi-Ne	2.Kharai		3.Niang Megha
1.Mewati		3.Nilgiri		3.Marwari		4. Nicobari
2. Ponwar			**Vulnerable**			
3.Punganur			1. Konkan Kanyal			
4.Siri		**Vulnerable**		**Not at risk**	**Not at risk**	
5.Vechur		1.Gurez	**Not at risk**	1.Bikaneri	1.Kathiawari	
Not at risk		2.Jalauni	Rest all breeds	2.Jaisalmeri	2.Marwari	
Rest all breeds		3.Kendrapada		3.Kutchi		
		4.Poonchi				
		5.Rampur Bushair				
		Not at risk				
		Rest all breeds				
Poultry (Chicken)						
Vulnerable						
1.Kalasthi						
Not at risk						
Rest all breeds						

Table 3: Breeds of Indian livestock at risk as per breed survey 2013

Species	Breed	Risk status category	Remarks for such status
Cattle	Bargur	Vulnerable	Breeding females are more than 3,000 but less than 6,000
	Krishna Valley	Endangered	Breeding females are more than 300 but less than 3,000
	Mewati	Endangered	Breeding males are more than five and less than 20 and rate of in breeding is high
	Pullikulam	Endangered	Breeding females are more than 300 but less than 3,000, breeding males are only 25 and rate of inbreeding is high
	Punganur	Endangered	Breeding females are more than 300 but less than 3,000
	Siri	**Vulnerable**	Breeding females are more than 3,000 but less than 6,000
	Vechur	**Critical**	**Breeding males are less than five, breeding females only 494 and rate of inbreeding is high**
Buffalo	Chilika	Endangered	Breeding females are more than 300 but less than 3,000, breeding males are only 25 and rate of inbreeding is high
	Toda	Endangered	Breeding females are more than 300 but less than 3,000
Goat	**Attapadi Black**	**Vulnerable**	Breeding females are more than 3,000 but less than 6,000
Sheep	Karnah	Endangered	Breeding females are more than 300 but less than 3,000
	Poonchi	Endangered	Breeding females are more than 300 but less than 3,000
Pig	Australian Large Black	Endangered	Breeding females are more than 100 but less than 1,000
	Duroc	Endangered	Breeding females are more than 100 but less than 1,000
	Saddleback	Endangered	Breeding females are more than 100 but less than 1,000

Double-humped camel (*Camelus bactrianus*): critically endangered since 1964

The word *bactrianus* in *Camelus bactrianus* comes from Bactria, a kingdom at the foot of the Hindu Kush Mountains of ancient Persia (Lensch, 1999). The Bactrian camels are mostly confined **in high land of Central Asian countries viz. Mongolia, China, Kazakhstan, Turkmenistan, North-Eastern Afghanistan, Uzbekistan and to lesser extent in Pakistan, Iran and Turkey** (He, 2002; Indra, *et al.*, 2003; Isani & Baloch, 2000; Moqaddam & Namaz-Zadah, 1988).

The number of double-humped camels revealed by door-to-door survey conducted by the Department of Animal Husbandry, Ladakh was 56 in 1964, 32 in 1978 and 47 in 1986 (Khanna & Khan 1988). NRC on camel Bikaner, conducted studies during an ICAR AP-Cess fund scheme on "Evaluation and conservation of double humped Camels in cold desert region." According to an initial survey during 1996 the population of double humped camels (*Camelus bactrianus*) was found to be **76 confined to the 4-5 villages of the Nubra valley of Ladakh, situated at an altitude of 10,000 to 12,000 ft**. There were around 150 camels in Nubra in 2012 and three years later, there were 211. The largest population of double humped camels is in the Hunder village followed by Sumoor, Diskit and Tigger (Vyas *et al.,* 2015). **As per latest Livestock Census 2019 the camel population in Jammu and Kashmir (J & K) is 470.** Since dromedary camels are not available in J & K so it is safe to presume that this data is for Bactrian camels. The IUCN (International Union for the Conservation of Nature) has declared the camel **critically endangered since 1998**, yet no serious measures have been adopted to conserve this species (Makhdoomi *et al*., 2013). As the ingenuity of "Bactrian Camel Safaris" grew in popularity, the villagers decided to register their enterprise. The local administration has helped camel owners to form camel unions. And thus the "Central Asia Camel Safari", a registered cooperative society, came into existence in 2009. The neighbouring villages namely Sumoor (situated on the other side of Nubra valley near Siachin base), Tigger, and Diskit also formed their own Bactrian Camel Cooperatives and Unions.

Conservation Versus Preservation of Domestic Animals

- The word conservation is closely related to the preservation of domestic animals.
- The **preservation covers** the continued maintenance of genetic variability. The preservation of a breed is required when it reaches an **endangered level of population or near to extinction**. The preservation and breed multiplication may also be required in case of individuals having unique traits or exceptional genetic merit.
- **Conservation includes the preservation along with upgradation (improvement) of the genetic potential and management of a breed for use in future**. Thus, the conservation covers both continued maintenance of genetic variability, improvement and sustainable utilization by exploiting the genetic variability.
- **The conservation can also be defined as the management of the biosphere of human use for benefits in present time together with maintaining its potential to meet the future needs.**

What is conservation?

Where conservation biology is often focused on preventing ongoing degradation, restoration ecology seeks to actively reverse such degradation. Conservation biology is a multidisciplinary science that assists conservation practitioners in addressing the loss of our biological resources.

What is preservation and restoration of the ecosystem?

Preservation is safeguarding and protecting the identified resource. Restoration on the other hand is the shift or repair of an ecosystem that is damaged or not functioning. Restoration involves returning a system to a former state or to a functioning state.

Restoration ecology is the scientific study of repairing disturbed ecosystems through human intervention. Where conservation biology is often focused on preventing ongoing degradation, restoration ecology seeks to actively reverse such degradation. Conservation biology is a multidisciplinary science that assists conservation practitioners in addressing the loss of our biological resources. Conservation biology has two central goals: to evaluate human impacts on biological diversity, and to develop practical approaches to prevent the extinction of species and maintain the integrity of ecosystems.

How are endangered species preserved?

The most common are creation of protected areas, captive breeding and reintroduction, conservation legislation, and increased public awareness.

Principles of Conservation

The conservation of **Animal Genetic Resources (AnGR)** should be based on the following basic principles

1. **Maintaining optimum population size:** The stock of animals should be maintained with optimum population size or number above the level of risk as much as possible.
2. **Characteristics of the stock:** The stock for preservation/conservation should be of the following types

   a. Pure form given most priority.

   b. Identifying special traits in a breed of interest.

   c. Maintain diverse stock for genetic security.

   d. Breed registration program.

3. **Environmental conditions**
   a. Maintain and conserve the locally adapted breeds in the same location.
   b. Maintain live animals providing the similar feeding, management and environmental conditions under which they had been traditionally kept.
   c. For endangered animals are given special housing management and feed to combat adverse climate.
4. **Breeding methods/Genetic improvement:** The genetic merit and diversity should be maintained using appropriate breeding programmes to improve genetic diversity.
5. Application of latest technology for breed multiplication like somatic cell cloning, IVF, MOET, Al etc.

Methods of Conservation of Animal Genetic Resources (AnGR)

- The live-stock biodiversity is essential for a sustainable and efficient production of food in order to satisfy the different needs of the wide variety of consumers, starting from the increasing demand for food and agriculture in the world.
- Genetic diversity is a prerequisite for genetic improvement and environmental adaptation.
- However, the diversity of Animal Genetic Resources (AnGR) has been in a continual state of decline as human population and economic pressures accelerate the rate of change in traditional agricultural systems.
- As a result, more and more breeds of domestic animals are in danger of becoming extinct.
- The World Watch List for Domestic Animal Diversity (WWL-DAD), is an information system (IS) to identify and monitor domestic animal breeds all over the world aiming to reduce the loss of biodiversity due to genetic erosion. A total of 6379 livestock breeds were reported belonging to 30 different species in WWL-DAD 3rd edition.
- Population size data was available for **4183 breeds of which 740 breeds were already extinct and 1335, or 32%, were classified at high risk of loss and were threatened by extinction**.
- The number of breeds reported constantly increased. However, in 2008, 48% of 10550 mammalian national breed populations and **53%**

of the 3450 avian national breed populations recorded in DAD-IS (Domestic Animal Diversity Information System), had no population data recorded; thus, the **risk-status categories of approximately 64%** of reported breeds were available in the DAD-IS.

The conservation of AnGR can be done *in situ* as well as *ex situ*, which are detailed below:

1. *In situ* Conservation

This involves the maintenance of the live population of animals in their adaptive environment (original and natural conditions) and animal population continues to evolve and be developed for more sustainable use. There are two aspects of in situ conservation viz., active and passive.

- The active in situ conservation is equivalent to **breed development** through animal breeding programme whereas the **passive in situ conservation is the maintenance of live animal populations within their environments for the breeds which are at risk of loss**. This method is best for **conserving a breed.**
- The **sufficient genetic variability in the breed population** can be maintained if the sample size is fairly large. A proper breeding plan for genetic improvement can effectively work to make the breed economically viable.
- Moreover, the breed can **gradually adapt to the changing environment over time**. However, the limiting factor for its implementation is the high cost involved in maintaining large herds/flocks.
- The ***in situ* conservation can be done in a better and scientific way at the organized farms under the control of Govt. to ensure purity of breed.**
- **Identifying distinct breed populations and describing their external and production characteristics within a given production environment.**
- **Nowadays, analyses at the DNA level are mainly based on high-throughput assays and technological platforms detecting thousands of genetic polymorphisms simultaneously.**
- **Molecular studies allow to identify and monitor the genetic diversity within and across breeds, to infer the population structure.**

- **Reconstruct evolution history, and to discover evidence on mankind history, as in the case of the novel clues found from *Bos taurus* mitochondrial DNA about the mystery of Etruscan origins.**
- **The basis of mitochondrial DNA validated the hypothesis of a common past migration of humans and cattle, reaching Etruria from the Eastern Mediterranean area by sea.**
- This method Is also feasible in field conditions i.e. in herds maintained by farmers and NGO like BAIF, AFPRO, Goshalas. The conservation can be done in field conditions by forming breed societies.

Institutional herds: Some important indigenous breeds are kept by 1CAR institutes. SAU's animal breeding farms. central and state Govt. livestock farms. However the population size is too small for improvement through selection. The conservation of breeds on organized farms together with improvement in their productivity can be done by the followings:

- Effective population size
- MOET
- Nuclear herd both ONBS and CNBS
- Improving the management and environment

The organized farms have a small number of breedable populations. It is therefore very important to know the minimum population size of a breed for in situ conservation so as to avoid the risk of inbreeding and genetic drift which are increased with the decrease in population size. The organized farms having a nucleus herd of 100 breedable females at least can run the selective breeding programme for It is, improvement and in situ conservation.

Farmers herds: The individual farmer maintains small herds/flocks. Therefore, they can not keep the male germplasm of good quality. This forces them to mate their females with the unselected bulls of low genetic merit available in and around the village. This results in production of non-descript populations.

Secondly, no data recording system is followed in farmer's herds/locks. These problems of maintaining the purebred population of indigenous breeds and data recording can be solved:

by forming breed societies for development, improvement and conservation of indigenous livestock breeds. The breed society may be constituted having the Distt Collector, Deputy Director AH District. Livestock Officer/VAS, Block Pramukh/ Pradhan, Farmers, Animal Breeders etc.

- by covering the area of breed under Herd Registration Scheme for data recording and
- by giving incentives to the farmers in different ways viz. an amount for keeping purebred animals, provision of semen and A.I. facilities and support price of products from purebreds etc.

The following strategies to increase fodder resources may be helpful for in situ conservation of indigenous breeds:

- To maintain the common grassland and other feed/fodder resources for razing of animals.
- To reduce the grazing pressure on depleted vegetation and feeding resources by improving the migrating grazing system during lean period (summer) in a way to move the animals to the forests, riverside wasteland and harvested agricultural fields and return to native place in the rainy season. It is also essential and mandatory for scientific management and seeding of waste Lands/grasslands and controlled grazing.
- Controlled grazing in forests because livestock and grassland ecology are complementary to each other for the overall maintenance of sustainable ecological balance.
- Planting the high yielding species of plants which provide optimal lopping in common grasslands and pastures, so as the livestock can be supplemented with tree lopping when grazing resources are depleted in lean period.
- This is known as silvopasture and multi-tier vegetation. Besides above strategics to increase the fodder resources, the following programmes will be helpful for in situ conservation:
- Improving economic efficiency of indigenous breeds through use of superior germplasm by selective breeding, better health coverage and providing balanced ration
- To create public awareness about the local breeds for their suitability and special attributes under nature conditions.

Genetic diversity and food security

- An example of the nutraceutical properties linked to animal food concern biopeptides which can be released from animal protein by the action of proteolytic enzymes during human digestion. Milk proteins represent a reservoir for a wide variety of bioactive peptides, which could affect milk nutritional value. The activity of biopeptides might be affected by amino-acid exchanges or deletions resulting from gene mutations and thus linked to particular breeds. Thus, the reason for preserving a particular breed can be related to the importance of conserving biological components of particular significance from a nutritional point of view.

- The conservation of livestock variability is also a crucial element in order to preserve and implicate specific nutritional and nutraceutical properties of animal products.
- The null alleles of particular casein genes, occurring at different frequencies depending on the goat breed, are an example of this important aspect, indicating that milk from goats carrying particular casein genotypes could be devoted to feeding children allergic to specific milk proteins, and that local breeds could be a reservoir of such genotypes.
- Milk from domestic cows has been an important food source for over 8000 years, especially in lactose-tolerant human societies. In fact, some individuals have the genetically determined ability to digest lactose by the action of persistent lactase enzymes in adulthood, thereby benefiting from the rich food resources occurring in cows' milk. Cow's milk protein genes and humane lactase genes underwent a gene-culture co-evolution in Europe resulting from the divergent use of milk (cheesemaking, direct consumption) in the different areas of Europe. This is impressive proof of the non-random occurrence of milk protein genetic variation over the centuries.
- "*In situ* conditions" were defined in the Convention on Biological Diversity as "conditions where genetic resources exist within ecosystems and natural habitats, and, in the case of domesticated or cultivated species, in the surroundings where they have developed their distinctive properties".
- It also includes the monitoring, characterization, and utilization over time of the gene pool of each species. Efficient in situ conservation strategies are based on the real self-sustainability of a breed. Since we cannot preserve each breed or population for economic reasons, we must decide what to safeguard (special character of products) on the basis of the real value of a breed.
- The link of a local breed to a typical product is a winning promotion tool for the breed. Further Italian examples are the Fatulì cheese made from raw milk of the Bionda dell'Adamello goat, the Cinta Senese ham and

sausages produced from the meat of this local pig of the Tuscany region, some dairy products from the milk of the Rendena cows reared in the Italian Alps, the Fiorentina of the Chianina, the well-known Fontina cheese linked to the Valdostana breeds in Valle d'Aosta Mountain region.

2. *Ex situ* Conservation

Ex situ conservation mainly refers to cryo-conservation of semen, ova, embryos or tissues. The Convention on Biological Diversity emphasizes the importance of in situ conservation and considers *ex situ* conservation as an essential complementary strategy to *in-situ.*

The *in-situ* conservation may not be feasible for the breeds which are economically not viable and hence they may be lost due to economic pressure. The *ex-situ* conservation can be done to handle this situation. The *ex-situ* conservation can be *in vivo* and *in vitro.*

In vivo method

It is ex situ **conservation of live animals in small numbers at a place away from the main breeding tract of the breed viz. organized herd of a research station, central/ state government animal farms, bull mother farms, zoo and breeding park**. This conservation has the type limitation in terms of population size to avoid the adverse effect of inbreeding. It is not possible to keep the large herds/flocks of breeds which are economically not viable. Inbreeding in a small population is unavoidable and hence the genetic defects appear. Therefore, the breeding population should be maintained in such a manner that the inbreeding is minimum and the performance of the population showed improvement over the years so as the breed becomes self-sustaining. The effective population size can be maintained on organized farms or with farmers in the native breeding tract or under livestock breed safari/parks on the pattern of wildlife.

The live animal reservoir/breed safari/park may help to preserve the natural habitat keeping all species of plants, animals and other organisms. It will be a good system to conserve the precious germplasm of breeds of domestic animals as a self-sustaining unit. This may also attract the tourist industry as in developed countries.

These may serve as the amusement or safari or zoological parks if managed properly keeping a diverse population of distinct breeds and may thus provide occasion for urban people to know about these endangered animals and their merits.

In vitro method

It is the storage of living cells for a long period of time. It is done by (1) deep freezing of sperms (semen) and oocytes, (i) deep freezing of embryos, and (iii) storage of DNA.

The in vitro conservation has the following basic objectives

- Regeneration of endangered breed
- New breed development
- Supporting the in vivo populations
- Research for determining the effect of single major gene
- DNA studies and genome mapping

(i) **Cryopreservation of sperms and oocytes:** The semen (sperms) of all the species is suitable for deep freezing. The technique of freezing, storing and thawing of semen is simple and a routine one. Likewise, the cryogenic storage of oocytes as a portion of ovaries from slaughtered animals can be done for most of the species of farm animals, except cattle. The conservation of the haploid genome as semen, but not the oocytes, has great limitations of the requirement, for IVF of the animals of opposite sex or both types of cells (sperms and oocytes) of the same breed.

(ii) **Cryopreservation of embryos:** It has been found successful to varying degree in case of cows, buffaloes, sheep, goats and horses. The embryo being a diploid zygote contains all the genetic information and hence it is an excellent tool of conservation. But it is relatively expensive to obtain embryos. Secondly, the cryo-preservation techniques need further refinement for economical use. It requires 8 to 12 embryos to be frozen to get one live animal. This can be used for conservation of at least endangered breeds. About 300 frozen embryos are required for conservation of a breed assuming the pregnancy rate of 40% and survival rate of 50%.

(ii) **Storage of DNA:** The genetic material can be preserved as the cryogenic storage of NDA. This method avoids the spreading of disease during its transportation. The storage of uncatalogued DNA is already possible but the genome maps of different species are not yet available for want of which, the storage of DNA could not become the normal method of preservation.

Recent/New approaches

Embryonic stem cells: These are obtained from culture of inner cell mass of a young blastocyst. These are known as totipotent (undifferentiated) cells which are cultured in vitro for a very long period. These cells can be multiplied indefinitely and can be obtained relatively easily from cultured young embryos or cultured cells (primordial germ cells) and are routinely kept frozen before use. The ceils have potential to develop into several embryos i.e. the single fertilized zygote gives rise to many embryos capable of being frozen for a very long time.

Somatic cells cloning: The live animal can be produced from stored somatic cells. The cloning produces a series of exact replicas/copies of the concerned animals from which the somatic cells have been used. Two techniques (Roslin and Honolulu) which are in use at present enucleate the egg cell to eliminate the majority of the information.The Honolulu method is quicker than Roslin as well as it does not require the in vitro culture of donor cells. In Roslin technique, the donor cell is forced to remain in the dormant cell stage and cultured in vitro to produce multiple copies of the same nucleus. The enucleated recipient cells and the nucleated donor cell are fused by electric pulse to develop an embryo. The surviving embryo is incubated in sheep's oviduct and finally placed in the uterus of surrogate mother ewe. In the Honolulu method, the unfertilized mouse egg cells are used as the recipient of donor nucleus. The egg cell, after acceptance in a new nucleus, is grown in a chemical culture. The developing embryo is transplanted in a surrogate mother.

***In vitro* fertilization (IVF), MOET can be used for rapid multiplication of endangered breeds.**

Ex situ is the method of choice to safeguard farm Animal Genetic Resources against disasters and is a valuable option when socio-economic, cultural and ecological values linked to a breed are already missing or of no interest.

Genetic materials cryopreserved in gene banks can be used for different purposes

1. to reconstruct a breed in case of extinction or loss of a substantial number of animals,
2. to create new lines or breeds,
3. as a back-up to quickly modify and/or reorient selection programmes,
4. to support populations conserved in vivo in cryo-aided live schemes, or

5. as a genetic resource for research.
6. Gene banks can also be used also for storing alleles of a particular locus that are being eradicated through artificial selection programs.
7. Concerning the support to the living population semen stored in gene banks for short-term storage can be distributed to farmers for the use in breeding programs of local breeds, whereas long-term storage of semen, called genetic reserve, can be considered as back up to secure the breed.

National Gene Banks

In order to rationalize cryo-storage of AnGR, some European countries have created **national gene banks**. These include: Austria (Austrian Gene Bank for Farm Animals), France (Cryobanque Nationale), Netherland (Centre for Genetic Resources—Gene Bank) and Nordic Countries (Denmark, Finland, Iceland, Norway and Swe- den: Viking Genetics, the Danish-Swedish-Finnish AI- centre).

The protection and preservation of biodiversity and ecosystems is one of the main objectives of the Sixth Environment Action Programme and requires economic and scientific efforts that span from biotechnologies for **germplasm conservation and genotyping to biobanking creation and to bioinformatics tool development for the data analysis and digital preservation**. However, the impressive increase of the technological tools as well as the valuable work of researchers, bioinformatics and other scientists, represents one of the major hopes for the preservation of biodiversity in the future.

ICAR- National Bureau of Animal Genetic Resources (*ICAR-NBAGR*)

Established on 21st September, 1984 at Bangalore in the form of twin institutes namely ICAR- National Bureau of Animal Genetic Resources and National Institute of Animal Genetics and then shifted to Karnal in 1985, the two institutes were merged to function as a single entity in the form of ICAR-National Bureau of Animal Genetic Resources (ICAR-NBAGR) in 1995. This premier institute is dedicated to work with its mandate of identification, evaluation, characterization, conservation and utilization of livestock and poultry genetic resources of the country.

NBAGR has registered ten new breeds of indigenous livestock species in the country during 2023, bringing the total number of registered indigenous breeds to 220.

In line with our commitment to assess the risk status of indigenous breeds, NBAGR introduced the Breed Watchlist-2022. This Watchlist serves as a valuable indicator for prioritizing the conservation and efficient management of AnGR. Out of a total of 38 indigenous breeds at risk, they have successfully cryopreserved 20 at the National Gene Bank of the institute.

Establishment of breed conservation centres in the respective native breed tracts for *in-situ* conservation and national gene bank, DNA bank, somatic cell bank for *ex-situ* conservation at the institute level is a remarkable step for conservation of economically important and environmentally well adapted breeds. The Bureau is regularly conducting various national and international trainings; demonstration/exhibition and sensitization programmes for the researchers, field functionaries, State AH officers and livestock keepers.

ICAR-NBAGR registered seven new breeds of indigenous livestock species and one synthetic cattle breed in the country. These breeds are – Aravali Chicken (Gujarat), Andamani Duck (Andaman and Nicobar Islands), Anjori Goat (Chhattisgarh); Andamani Goat (Andaman and Nicobar Islands); Bhimthadi Horse (Maharashtra), Andamani Pig (Andaman and Nicobar Islands), Macherla Sheep (Andhra Pradesh), and Frieswal Cattle (Uttar Pradesh and Uttarakhand). After including these breeds, total number of registered breeds has been reached to 220, including 53 for cattle, 20 for buffalo, 39 for goat, 45 for sheep, 8 for horses & ponies, 9 for camel, 14 for pig, 3 for donkey, 3 for dog, 1 for yak, 20 for chicken, 3 for duck, 1 for geese and 1 for synthetic cattle. The Bureau also allotted Accession numbers to these newly registered breeds.

Milk Protein Variants Another Potential Need of Conservation of Indigenous Livestock Species

1. A1 milk versus A2 milk

- The above **A1** type of **milk** is mostly produced by European cow breeds such as Holstein Friesian (HF), Ayrshire and British Shorthorn.
- On the other hand, Jersey and Guernsey cows in the Channel Islands, the Charolais and Limousin breeds of Southern France, and the **Zebu cattle** of Africa and Asia produce **A2 milk**, which does not release BCM-7.
- BCM-7 is hypothesized to interact with μ-opioid receptors on immune cells in humans.
- Non-cow milk, including that of **humans, sheep, goats, donkeys, yaks, camels, buffalo, and others, also contain mostly A2 β-casein**, and so the term "A2 milk" is also used in that context.
- Regular milk contains both A1 and A2 beta-casein, but A2 milk contains only A2 beta-casein.

- Some studies suggest that A1 beta-casein may be harmful and that A2 beta-casein is a safer choice.
- Caseins constitute the vast majority of the proteins in milk. Among these, β-casein comprises around 37% of all caseins, and it is an important type of casein with several different variants.
- The A1 and A2 variants of β-casein are the most researched genotypes due to the changes in their composition. It is accepted that the A2 variant is ancestral, while a point mutation in the 67th amino acid created the A1 variant.
- The opioid-like characteristics of BCM-7 are highlighted for their potential triggering effect on several diseases. Most research has been focused on gastrointestinal-related diseases; however other metabolic and nervous system-based diseases are also potentially triggered. By manipulating the mechanisms of these diseases, BCM-7 can induce certain situations, such as conformational changes, reduction in protein activity, and the creation of undesired activity in the biological system.
- Furthermore, the genotype of casein can also play a role in bone health, such as altering fracture rates, and calcium contents can change the characteristics of dietary products. The context between opioid molecules and BCM-7 points to a potential triggering mechanism for the central nervous system and other metabolic diseases.
- Individuals with lactose intolerance prefer A2 milk to conventional A1 milk, as BCM-7 in A1 milk can lead to inflammation and discomfort in sensitive individuals. A2 milk, which contains A2 β-casein, is believed to be more easily digestible than A1 β-casein. Its popularity has grown owing to reports linking A1 casein to diseases such as type 1 diabetes, heart disease, and autism. A2 milk has gained popularity as an alternative to A1 milk, primarily because of its potential benefits for individuals with certain diseases.
- The difference between A1 and A2 goat milk is that goat milk contains mostly A2 casein, whereas regular cow's milk contains a varying mixture of both A1 and A2 beta-casein.

2. Benefits of goat milk

Goat milk has several advantages over cow milk, making it a popular choice for some people. Here are a few reasons why goat milk might be considered better:

Easier to Digest: Goat milk contains less lactose than cow milk, which can make it easier to digest for people with lactose intolerance. Additionally, the fat globules in goat milk are smaller, which can aid in digestion.

Nutrient-Rich: Goat milk is higher in certain nutrients like calcium, vitamin A, and potassium compared to cow milk. It also contains more medium-chain fatty acids, which are easier for the body to metabolize.

Less Allergenic: Goat milk has a different protein structure than cow milk, specifically it **contains less of the protein alpha-S1-casein, which is often responsible for milk allergies**.

Better for the Environment: Goats generally require less space and food compared to cows, which can make goat milk production more sustainable.

References

NBAGR database http://14.139.252.116:8080/appangr/openagr.htm

DAHD website database https://dahd.nic.in/

Dairy Knowledge Portal https://www.dairyknowledge.in/

FAO database https://www.fao.org/dairy-production-products/production/dairy-animals/cattle/en/

Wikipedia.org web base encyclopaedia

Handbook of Animal Genetics and Breeding by SS Tomar

Instruction Manual for Integrated Sample Survey, provided by AHS Division, DADF, Ministry of Agriculture, GOI

"National Dairy Development Board". https://www.nddb.coop/

14

Livestock Census of India

20th Livestock Census of India

India has highest cattle population (192.49 million), buffalo population (109.85), in the world producing 24% is the highest milk yield (221.1 million tons in the year 2021-2022, and per capita availability was 444 gm/day) in the world (Source NDDB). India produces 43% of world buffalo meat production. Unfortunately, 80% cattle and 70% buffaloes are non-descriptive and yield very low amount of milk.

After independence Indian Council of Agricultural Research (ICAR) has developed serval projects and plans for research and characterization of livestock species all over the country. Our country holds enormous genetic biodiversity in our domestic animals, result in increasing number of registered breeds in every decade. Breed registration Committee on its online meeting held on 5th December, 2023 approved registration of eight new breeds of livestock. This includes Aravali Chicken (Gujarat) Andamani Duck (Andaman and Nicobar Islands), Anjori Goat (Chhattisgarh), Andamani Goat (Andaman and Nicobar Islands), Bhimthadi Horse (Maharashtra), Andamani Pig (Andaman and Nicobar Islands), Machcrla Sheep (Andhra Pradesh), Frieswal Cattle (Uttar Pradesh and Uttarakhand). Endangered, critical and vulnerable breeds of livestock also have been identified by ICAR which need more scientific research and multiplication breed.

After including 8 new breeds, total number of registered breeds has been reached to 220, including 53 for cattle, 20 for buffalo, 39 for goat, 45 for sheep, 8 for horses & ponies, 9 for camel, 14 for pig, 3 for donkey, 3 for dog, 1 for yak, 20 for chicken, 3 for duck, 1 for geese and 1 for synthetic cattle. Bureau also allotted Accession numbers to these newly registered breeds.

Department of Animal Husbandry & Dairying releases 20th Livestock Census; Total Livestock population increases 4.6% over Census-2012, Increases to 535.78 million. The Census will prove beneficial not just for policy makers but also for agriculturists, traders, entrepreneurs, dairying industry and masses in general. This release provides some key results reflecting the aggregate counts of various species as well as its comparison with previous census. The 20th Livestock Census was carried out in about 6.6 lakhs villages and 89 thousand urban wards across the country covering more than 27 Crores of Households and Non-Households.

Key Features

Some of the key features of the 20th Livestock Census are given below:

1) The total livestock population is 535.78 million in the country showing an increase of 4.6% over Livestock Census 2012

2) **Total bovine population (cattle, buffalo, mithun and yak) was 302.79 million in 2019, (highest in the World), which showed an increase of 1.0% over the previous census.**

3) The total cattle population in the country was 192.49 million in 2019 showed an increase of 0.8 % over previous census.

4) The female cattle (cows population) was 145.12 million, which increased by 18.0% over the previous census (2012).

5) The exotic/crossbred and indigenous/non-descript cattle population in the country was 50.42 million and 142.11 million, respectively.

6) The **indigenous/non-descript female cattle** population has increased by 10% in 2019 as compared to previous census.

7) The population of exotic/crossbred cattle has increased by 26.9 % in 2019 as compared to previous census.

8) **There is a decline of 6% in the total indigenous** (both descript and non-descript) cattle population over the previous census. However, the pace of decline of indigenous cattle population during 2012-2019 was much lesser as compared to 2007-12, which was about 9%.

9) The total buffaloes in the country were 109.85 million showing an increase of about 1.0% over previous Census.

10) **The total milch animals (in-milk and dry) in cows and buffaloes were 125.34 million, an increase of 6.0 % over the previous census.**

11) The total sheep in the country is 74.26 million in 2019, increased by 14.1% over previous Census. (**2nd in the World**)

12) The Goat population in the country in 2019 is 148.88 million showing an increase of 10.1% over the previous census. (**Ist in the World**)

13) The total Pigs in the country is 9.06 million in the current Census, declined by 12.03% over the previous Census. (5th **in the World**)

14) The total Mithun in the country is 3.9 Lakhs in 2019, increased by 30.0% over previous Census.

15) The total Yak in the country is Fifty-Eight Thousand in 2019, **decreased by 24.67%** over previous Census.

16) The total Horses and Ponies in the country is 3.4 Lakhs in 2019, **decreased by 45.6%** over previous Census.

17) The total population of Mules in the country is Eighty-Four Thousand in 2019**, decreased by 57.1% over previous Census.**

18) The total population of Donkeys in the country is 1.2 Lakhs in 2019, **decreased by 61.23% over previous Census.**

19) The total Camel population in the country is 2.5 Lakhs in 2019, **decreased by 37.1% over** previous Census.

20) The total Poultry in the country is 851.81 million in 2019, increased by 16.8% over previous Census.

21) The total Backyard Poultry in the country is 317.07 million in 2019, increased by 45.8% over previous Census.

22) The total Commercial Poultry in the country is 534.74 million in 2019, increased by 4.5% over previous Census.

23) Distribution of Livestock Population

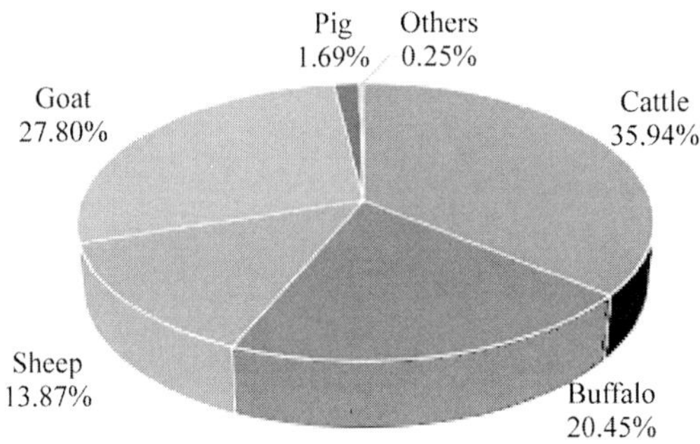

Graph 1: Livestock Population 2019 - Share of Major Species

24) In 20th Livestock Census, 35.94% -Cattle, 27.80%- Goat, 20.45%- Buffaloes, 13.87%-Sheep, 1.69%-Pigs.

25) Mithun, Yaks, Horses, Ponies, Mules, Donkeys and Camels taken together contribute 0.23% of the total livestock.

26) As compare to previous census the percentage share of sheep and goat population has increased whereas the percentage share of cattle, buffalo and pig has marginally declined.

27) Livestock Population

Category	Population (in million) 2012	Population (in million) 2019	% growth
Cattle	**190.9**	**192.49**	**0.83**
Buffalo	**108.7**	**109.85**	**1.06**
Sheep	65.07	74.26	14.13
Goat	**135.17**	**148.88**	**10.14**
Pig	10.29	9.06	-12.03
Mithun	0.3	0.38	26.66
Yak	0.08	0.06	-25
Horses & Ponies	0.63	0.34	-45.58
Mule	0.2	0.08	-57.09
Donkey	0.32	0.12	-61.23
Camel	0.4	0.25	-37.05
Total livestock	**512.06**	**535.78**	**4.63**

Livestock population of major species during 2007 to 2019

Species	Population (in million) 2007	Population (in million) 2012	Population (in million) 2019
Cattle	199.08	190.9	192.49
Buffaloes	105.34	108.7	109.85
Sheep	71.56	65.07	74.26
Goats	14.054	135.17	148.88
Pigs	11.13	10.29	9.06
Mithun	0.26	0.3	0.38
Yaks	0.08	0.08	0.06
Horses & ponies	0.61	0.63	0.34
Mules	0.14	0.2	0.08
Donkeys	0.44	0.32	0.12
Camels	0.52	0.4	0.25
Total livestock	**529.7**	**512.06**	**535.78**

Livestock population, 2012 and 2019 of major states

S. No.	States	Population (in million) 2012	Population (in million) 2019	% Change
1	Uttar Pradesh	68.7	67.8	-1.35
2	Rajasthan	87.7	56.8	-1.66
3	Madhya Pradesh	36.3	40.6	11.81
4	West Bengal	30.3	37.4	23.32
5	Bihar	32.9	36.5	10.67
6	Andhra Pradesh	29.4	34	15.79
7	Maharashtra	32.5	33	1.61
8	Telangana	26.7	32.6	22.21
9	Karnataka	27.7	29	4.7
10	Gujarat	27.1	26.9	-0.95

Poultry population in India from 2012 and 2019

	Population (in million) 2012	Population (in million) 2019	% growth
Total poultry	729.21	851.81	16.81
Backyard poultry	217.49	317.07	45.78
Commercial poultry	511.72	534.74	4.5

Poultry Population, 2012 & 2019 of Major State

Sl. No.	States	Population (In million) 2012	Population (In million) 2019	% Change
1	Tamil Nadu	117.3	120.8	2.92
2	Andhra Pradesh	80.6	107.9	33.85
3	Telangana	80.8	80.0	-0.93
4	West Bengal	52.8	77.3	46.34
5	Maharashtra	77.8	74.3	46.34
6	Karnataka	53.4	59.5	11.33
7	Assam	27.2	46.7	71.63
8	Haryana	42.8	46.3	8.11
9	Kerala	24.3	29.8	22.61
10	Odisha	19.9	27.4	37.95

Table 1: Breed wise number of animals under indigenous cattle (20th Livestock Census, 2019)

Sl. No.	Breed Name	Pure (No.)	Graded (No.)	Total (No.)	Percentage share w.r.t total
1	Gir	23,00,090	45,57,694	68,57,784	4.8
2	Lakhimi	66,48,519	1,80,965	68,29,484	4.8
3	Sahiwal	18,81,453	40,68,221	59,49,674	4.2
4	Bachaur	32,15,259	11,30,681	43,45,940	3.1
5	Hariana	11,79,089	15,78,097	27,57,186	1.9
6	Kankrej	15,80,802	6,34,735	22,15,537	1.6

Sl. No.	Breed Name	Pure (No.)	Graded (No.)	Total (No.)	Percentage share w.r.t total
7	Kosali	9,89,803	5,66,871	15,56,674	1.1
8	Khillar	8,44,400	4,54,796	12,99,196	0.9
9	Rathi	8,78,852	2,90,976	11,69,828	0.8
10	Malvi	5,95,658	4,37,310	10,32,968	0.7
11	Badri	7,41,324	2,48,051	9,89,375	0.7
12	Hallikar	5,01,057	3,18,080	8,19,137	0.6
13	Malnad Gidda	6,31,530	81,528	7,13,058	0.5
14	Ongole	3,03,817	3,99,325	7,03,142	0.5
15	Red Sindhi	2,72,850	3,40,050	6,12,900	0.4
16	Tharparkar	1,51,056	4,31,201	5,82,257	0.4
17	Gangatiri	2,43,153	2,71,392	5,14,545	0.4
18	Nimari	3,98,341	8,0,720	4,79,061	0.3
19	Nagori	2,10,012	1,37,069	3,47,081	0.2
20	Amritmahal	1,04,990	1,96,364	3,01,354	0.2
21	Deoni	1,83,656	1,00,686	2,84,342	0.2
22	Motu	2,31,954	1,683	2,33,637	0.2
23	Dangi	1,39,971	51,724	1,91,695	0.1
24	Gaolao	1,12,563	74,324	1,86,887	0.1
25	Kenkatha	76,663	89,604	1,66,267	0.1
26	Kangayam	1,27,577	24,966	1,52,543	0.1
27	Red Kandhari	95,304	53,917	1,49,221	0.1
28	Binjharpuri	69,406	14,443	83,849	0.1
29	Bargur	42,300	13,159	55,459	-
30	Kherigarh	28,433	20,285	48,718	-
31	Ghumusari	35,626	3,261	38,887	-
32	Umblachery	31,195	11,195	42,390	-
33	Ponwar	14,480	14,176	28,656	-
34	Khariar	13,365	11,656	25,021	-
35	Siri	15,278	8,789	24,067	-
36	Krishna Valley	2,594	19,938	22,532	-
37	Mewati	9,024	12,877	21,901	-
38	Vechur	8,963	6,218	15,181	-
39	Pulikulam	10,495	3,439	13,934	-
40	Punganur	9,876	3,399	13,275	-
41	Belahi	4,238	1,026	5,264	-
Indigenous cattle breed		**2,49,35,016**	**1,69,44,891**	**4,18,79,907**	**29.5**
Nondescript			**10,02,26,559**		**70.5**
Total indigenous cattle				**14,21,06,466**	

Table 2: Breed wise number of indigenous buffaloes (20th Livestock Census, 2019)

Sl. No.	Name of the breed	Pure (No.)	Graded (No.)	Total no. of animals (No.)	Percentage share
1	Murrah	1,42,46,525	3,28,18,923	4,70,65,448	42.8
2	Mehsana	34,42,006	9,36,782	43,78,788	4.0
3	Surti	11,22,735	13,29,627	24,52,362	2.2
4	Jaffarabadi	11,13,789	10,25,338	21,39,127	1.9
5	Bhadawari	10,65,485	9,16,367	19,81,852	1.8
6	Banni	5,12,851	2,65,615	7,78,466	0.7
7	Pandharpuri	3,76,182	1,38,785	5,14,967	0.5
8	Marathwadi	2,17,581	25,069	2,42,650	0.2
9	Nili Ravi	1,08,659	1,27,590	2,36,249	0.2
10	Nagpuri	1,04,016	52,231	1,56,247	0.1
11	Kalahandi	18,123	7,041	25,164	-
12	Toda	14,497	5,371	19,868	-
13	Chilika	11,010	2,648	13,658	-
Indigenous buffalo		**2,23,53,459**	**3,76,51,387**	**6,00,04,846**	**54.6**
Non-descript		-	-	**4,98,46,832**	**45.4**
Total buffaloes		**2,23,53,459**	**3,76,51,387**	**10,98,51,678**	

Table 3: Breed wise number of animals under indigenous sheep (20th Livestock Census, 2019)

Sl. No.	Name of breed	Pure (No.)	Graded (No.)	Total (No.)	Percentage share with respect to Total
1	Nellore	77,89,472	62,54,363	1,40,43,835	20.0
2	Bellary	26,52,379	16,22,839	42,75,218	6.1
3	Marwari	17,46,966	11,23,091	28,70,057	4.1
4	Deccani	12,24,660	11,59,272	23,83,932	3.4
5	Kenguri	7,55,778	5,30,506	12,86,284	1.8
6	Mercheri	10,04,976	2,37,766	12,42,742	1.8
7	Patanwadi	5,05,595	3,74,025	8,79,620	1.3
8	Hassan	6,04,412	1,32,989	7,37,401	1.1
9	Jaisalmeri	4,55,956	2,24,217	6,80,173	1.0
10	Gaddi	3,39,799	3,27,116	6,66,915	1.0
11	Ramnad White	1,71,748	2,25,108	3,96,856	0.6
12	Chokla	2,25,304	1,56,893	3,82,197	0.5
13	Chottanagpuri	2,26,974	88,095	3,15,069	0.4
14	Madras Red	2,76,284	21,222	2,97,506	0.4
15	Nali	1,38,767	1,11,576	2,50,343	0.4
16	Mandya	1,15,973	1,34,065	2,50,038	0.4
17	Malapura	85,443	1,24,091	2,09,534	0.3
18	Bhakarwal	1,19,337	68,630	1,87,967	0.3
19	Pugal	1,40,482	29,968	1,70,450	0.2
20	Magra	86,113	45,576	1,31,689	0.2

Sl. No.	Name of breed	Pure (No.)	Graded (No.)	Total (No.)	Percentage share with respect to Total
21	Vembur	87,758	14,376	1,02,134	0.1
22	Coimbatore	38,960	61,357	1,00,317	0.1
23	Balangir	54,982	25,453	80,435	0.1
24	Garole	61,805	11,777	73,582	0.1
25	Changthangi	71,632	269	71,901	0.1
26	Bonpala	38,634	30,197	68,831	0.1
27	Chevaadu	55,348	12,525	67,873	0.1
28	Ganjam	50,905	10,574	61,479	0.1
29	Shahbadi	34,452	25,843	60,295	0.1
30	Kilakarsal	39,549	6,680	46,229	0.1
31	Jalauni	12,149	30,782	42,931	0.1
32	Muzzafarnagri	18,762	22,998	41,760	0.1
33	Sonadi	20,042	19,331	39,373	0.1
34	Gurez	6,970	18,744	25,714	-
35	Tiruchi Black	19,662	1,943	21,605	-
36	Poonchi	6,795	13,437	20,232	-
37	Rampur Bushair	8,894	9,345	18,239	-
38	Kendrapadda	8,293	2,681	10,974	-
39	Nilgiri	1,810	1,334	3,144	-
40	Karnah	1,430	1,691	3,121	-
41	Katchaikatty Black	1,506	394	1,900	-
42	Tibetan	317	1	318	-
Total indigenous sheep breeds		**1,93,07,073**	**1,33,13,140**	**3,26,20,213**	**46.5**
Non-descript		-	-	**3,75,52,269**	**53.5**
Total indigenous sheep		**1,93,07,073**	**1,33,13,140**	**7,01,72,482**	

Table 4: Breed wise estimated number of animals under indigenous goat (20th Livestock Census, 2019)

Sl. No.	Name of the breed	Pure (No.)	Graded (No.)	Total no. of animals (No.)	Percentage share
1	Black Bengal	2,52,77,770	23,84,206	2,76,61,976	18.6
2	Marwari	32,09,933	18,31,843	50,41,776	3.4
3	Barbari	20,30,278	27,29,027	47,59,305	3.2
4	Osmanabadi	23,91,392	12,05,679	35,97,071	2.4
5	Jamnapari	11,77,948	13,78,017	25,55,965	1.7
6	Sirohi	8,16,722	11,35,394	19,52,116	1.3
7	Kanni Adu	12,72,974	1,72,614	14,45,588	1.0
8	Beetal	5,71,474	6,63,286	12,34,760	0.8
9	Malabari	7,81,454	3,22,851	11,04,305	0.7
10	Gaddi	4,57,312	2,81,113	7,38,425	0.5

Sl. No.	Name of the breed	Pure (No.)	Graded (No.)	Total no. of animals (No.)	Percentage share
11	Jakhrana	2,89,191	3,66,391	6,55,582	0.4
12	Kutchi	4,41,547	1,42,991	5,84,538	0.4
13	Salem Black	4,05,600	86,392	4,91,992	0.3
14	Mehsana	2,61,039	1,61,470	4,22,509	0.3
15	Zalawadi	1,83,217	2,25,233	4,08,450	0.3
16	Kodi Adu	3,53,236	46,688	3,99,924	0.3
17	Gohilwadi	1,28,809	1,59,644	2,88,453	0.2
18	Surti	1,68,065	63,129	2,31,194	0.2
19	Ganjam	1,62,352	49,126	2,11,478	0.1
20	Changthangi	2,03,867	2,073	2,05,940	0.1
21	Sanganeri	1,27,343	35,748	1,63,091	0.1
22	Berari	65,927	18,896	84,823	0.1
23	Attapady Black	22,140	9,042	31,182	-
24	Pantja	19,576	9,152	28,728	-
25	Konkan Kanyal	9,628	7,264	16,892	-
26	Teressa	8,88	2,474	3,362	-
27	Chegu	1,316	1,040	2,356	-
28	Sumi- Ne	1,432	77	1,509	-
Indigenous goats		**4,08,32,430**	**1,34,90,860**	**5,43,23,290**	**36.5**
Non-descript		-	-	**9,45,61,496**	**63.5**
Total goat		**4,08,32,430**	**1,34,90,860**	**14,88,84,786**	

Table 5: Breed wise number of indigenous pigs (20th Livestock Census, 2019)

Sl. No.	Breed name	Total animals (No.)	Percentage share with respect to total
1	Doom	2,81,472	3.9
2	Niang Megha	2,21,962	3.1
3	Ghungroo	2,08,751	2.9
4	Nicobari	23,695	0.3
5	Tenyi Vo	4,471	0.1
6	Agnoda Goan	1,325	-
Indigenous pigs		**7,41,676**	**10.4**
Non-descript		**64,16,868**	**89.6**
Total indigenous pigs		**71,58,544**	

Table 6: Breed wise number of horses and ponies (20th Livestock Census, 2019)

Sl. No.	Name of the breed	Number of horses	Number of ponies	Total	Percentage share with respect to the total
1	Marwari	33,267	431	33,698	9.8
2	Kathiawari	23,941	1,339	25,280	7.4
3	Zanskari	6,314	346	6,660	1.9
4	Kachchhi-Sindhi	3,926	17	3,943	1.2
5	Manipuri	2,558	994	3,552	1.0
6	Bhutia	2,316	24	2,340	0.7
7	Spiti	1,064	66	1,130	0.3
Total breeds		**73,386**	**3,217**	**76,603**	**22.4**
Non-Descript		**2,24,798**	**40,825**	**2,65,623**	**77.6**
Total Horses & Ponies		**2,98,184**	**44,042**	**3,42,226**	

Table 7: Number of mules (20th Livestock Census, 2019)

Category	Mules (No.)
Under 3 years	17,003
3 years and above	67,258
Total	**84,261**

Table 8: Number of donkeys (20th Livestock Census, 2019)

Category	Spiti (No.)	Non-descript (No.)	Total (No.)
Male	4,203	60,539	64,742
Female	6,032	52,813	58,845
Total (Male + Female)	**10,235**	**1,13,352**	**1,23,587**
Percentage share with respect to their total (%)	**8.3**	**91.7**	

Table 9: Number of camels (20th Livestock Census, 2019)

Sl. No.	Name of the Breed	Number of Camels	Percentage share with respect to total
1	Bikaneri	74,551	29.6
2	Jaisalmeri	47,975	19.0
3	Kachchhi/Kutchi	16,905	6.7
4	Jalori	5,023	2.0
5	Marwari	2,720	1.1
6	Kharai	4,266	1.7
7	Mewari	465	0.2
8	Mewati	92	-
9	Malvi	102	-
Total Breeds		**1,52,099**	**60.4**
Non-descript		**99,857**	**39.6**
Total camels		**2,51,956**	

Table 10: Breed wise total number of fowls under backyard poultry and poultry farm (20th Livestock Census, 2019)

Sl. No.	Breed name	Backyard poultry	Poultry farm	Total	Percentage share with respect to total
1	Aseel	2,61,37,632	75,42,951	3,36,80,583	4.2
2	Miri	54,03,648	7,445	54,11,093	0.7
3	Daothigir	36,94,955	1,302	36,96,257	0.5
4	Kadaknath	15,99,417	4,38,058	20,37,475	0.3
5	Punjab Brown	1,68,859	11,96,941	13,65,800	0.2
6	Ankleshwar	7,02,147	6,19,859	13,22,006	0.2
7	Chittagong	23,855	12,17,145	12,41,000	0.2
8	Busra	9,95,386	2,23,338	12,18,724	0.2
9	Ghagus	4,55,230	3,05,734	7,60,964	0.1
10	Kashmir Favorolla	5,97,619	7	5,97,626	0.1
11	Haringhata Black	4,54,979	781	4,55,760	0.1
12	Danki	1,07,648	60,684	1,68,332	-
13	Tellichery	1,47,438	1,267	1,48,705	-
14	Kaunayen	1,10,706	4,122	1,14,828	-
15	Mewari	1,10,041	4,583	1,14,624	-
16	Hasli	85,733	5,652	91,385	-
17	Kalahasthi	11,736	38,000	49,736	-
18	Nicobari	25,641	974	26,615	-
	Total Desi Fowl	**408,32,670**	**1,16,68,843**	**5,25,01,513**	**6.5**
	Other Desi Fowl	**1867,31,428**	**5,31,78,591**	**23,99,10,019**	**29.7**
	Improved fowl	**532,09,966**	**46,22,31,485**	**51,54,41,451**	**63.8**
	Total fowl	**28,07,74,064**	**52,70,78,919**	**80,78,52,983**	

Table 11: Total number of turkey (20th Livestock Census, 2019)

Categories	Backyard poultry	Poultry farm	Total	Percentage share with respect to total
Male turkey	1,60,574	5,299	1,65,873	37.0
Female turkey	2,62,651	19,698	2,82,349	63.0
Total turkey	**4,23,225**	**24,997**	**4,48,222**	

Table 12: Total number of other poultry birds (20th Livestock Census, 2019)

Categories	Backyard poultry	poultry Farm	Total	Percentage share with respect to total
Quails	19,32,270	48,86,382	68,18,652	68.5
Other poultry birds	14,02,737	17,33,663	31,36,400	31.5
Total Others Poultry	**33,35,007**	**66,20,045**	**99,55,052**	

Table 13: Number of Mithun (20th Livestock Census, 2019)

Category	Male (No.)	Female (No.)	Total (No.)
Below three years of age	78,644	89,963	1,68,607
Three years and above	93,995	1,23,703	2,17,698
Total	**1,72,639**	**2,13,666**	**3,86,305**

Table 14: Number of Yak (20th Livestock Census, 2019)

Category	Male (No.)	Female (No.)	Total (No.)
Below three years of age	8,744	10,942	19,686
Three years and above	17,275	20,609	37,884
Total	**26,019**	**31,551**	**57,570**

Table 15: Number of Dogs (20th Livestock Census, 2019)

Male (No.)	Female (No.)	Total (No.)
70,58,379	23,75,660	94,34,039

Table 16: Number of Rabbits (20th Livestock Census, 2019)

Male (No.)	Female (No.)	Total (No.)
2,42,162	30,779	5,49,941

Table 17: Number of Elephants (20th Livestock Census, 2019)

Male (No.)	Female (No.)	Total (No.)
876	804	1,767

Milk, eggs and meat productions in India

Recently, the Ministry of Fisheries, Animal Husbandry and Dairying has released the 'Basic Animal Husbandry Statistics 2022, showing an increase in the milk, eggs and meat productions in India. The contribution of livestock in the agriculture sector has been showing steady improvement that signifies its growing importance for the country's economy.

What are the Key Highlights?

Milk production

Total milk production in India was 221.06 million tonnes in 2021-2022, being the highest **milk producing country in the world.** The milk production of the country increased by 5.29% over the previous year.

Indigenous cattle contribute 10.35% of the total milk production in the country, whereas non-descript cattle contribute 9.82%. However, the contribution of non-descript buffaloes to the total milk production in the countryis higher than that of cattle (13.49%).

In India, the proportion of milk production for different animals in 2024 is as follows:

Cows: Contribute approximately 47.3% of the total milk production.

Buffaloes: Being major contributor, account for around 49.0% of the total milk production

Goats: Contribute a smaller share, approximately 3.0%, to the overall milk production

If we see the share of major states in the total milk production, of the country, the top five major milk producing states are Rajasthan (15.05%), Uttar Pradesh (14.93%), Madhya Pradesh (8.06%), Gujarat (7.56%) and Andhra Pradesh (6.97%).

We should have proud that India is the largest producer of milk and buffalo meat, the 2nd largest producer of goat meat, 3rd in egg production and the 8th largest in overall meat production in the world.

Egg Production

The total egg production was 129.60 billion numbers, which increased by 6.19% than that of previous year.

Top five egg producing states are Andhra Pradesh (20.41%), Tamil Nadu (16.08%), Telangana (12.86%), West Bengal (8.84%) and Karnataka (6.38%). Here it is to mention that, these states together contribute 64.56% of total egg production in the country.

Meat Production

The total meat production in the country was 9.29 million tonnes, which increased by 5.62% as compared to the previous year.

The contribution to the total meat production from poultry is about 51.44%. The top five meat producing states are Maharashtra (12.25%), Uttar Pradesh (12.14%), West Bengal (11.63%), Andhra Pradesh (11.04%) and Telangana (10.82%). All together these states contribute 57.86% to total meat production in the country.

Wool

The total wool production in the country during 2021-22 was 33.13 thousand tonnes, which had declined by 10.30% as compared to previous year.

The top five major wool producing states are Rajasthan (45.91%), Jammu and Kashmir (23.19%), Gujarat (6.12%), Maharashtra (4.78%) and Himachal Pradesh (4.33%).

15

Question Bank for Animal Breeds

Q1. Fill in the blanks with appropriate answers.

1. India ranks ...**first**.... position in cattle population in the world.
2. India ranks ... **first** .... position in buffalo population in the world.
3. India ranks ... **first** .... position in total milk production in the world.
4. ...**Buffalo**........ species produce the highest amount of milk in India.
5. Frieswal cattle is a cross of ...**Holstein Friesian**....and ......**Sahiwal**...
6. Mehsana buffalo is a cross of ...**Murrah....and ......Surti**......
7. India ranks ... **first** .... position in goat population in the world.
8. India ranks ......**third**....... position in egg production in the world.
9. India ranks ...... **first** .... position in buffalo meat production in the world.
10. India ranks ......**8th**.... position in total meat production in the world.
11. The state ......**Rajasthan**.... has the highest goat population in India (2019).
12. The native home tract of Sahiwal breeds is ... **Montgomery** .. district of Pakistan.
13. The native home tract of Red Sindi breeds is ....**Sindh Pradesh**...... of Pakistan.
14. Total number of registered breeds in India till December 2023 by NBAGR is ......220....
15. Total number of registered cattle breeds in India till 2023 by NBAGR is ......53....
16. New registered synthetic breed of cattle by NBAGR is ... Frieswal Cattle...........

17. Total number of registered buffalo breeds in India till 2023 by NBAGR is ...20......
18. Total number of registered sheep breeds in India till 2023 by NBAGR is ...45.......
19. Total number of registered goat breeds in India till 2023 by NBAGR is ......39.......
20. Total number of registered horses and ponies in India till 2023 by NBAGR is ......8.......
21. Total number of registered camel breeds in India till 2023 by NBAGR is ......9.......
22. Total number of registered poultry breeds in India till 2023 by NBAGR is ......20.......
23. Total number of registered pig breeds in India till 2023 by NBAGR is ......14.......
24. Total number of registered duck breeds in India till 2023 by NBAGR is ......3.......
25. Total number of registered donkey breeds in India till 2023 by NBAGR is ......3.......
26. Total number of registered dog breeds in India till 2023 by NBAGR is ......3.......
27. Total number of registered geese breeds in India till 2023 by NBAGR is ......1.......
28. Total number of registered yak breeds in India till 2023 by NBAGR is ......1.......
29. Lola (loose skin), Lambi Bar, Montgomery, Multani are different names of ...**Sahiwal**....
30. Gir cattle originated at ....**Gir forests of South Kathiawar** ....of Gujrat.
31. Gir breed of cattle share ....**highest(4.8%)**......... among all breeds in India.
32. ...**Tharparkar** ...cattle have been derived from the place of its origin i.e. the Thar desert.

33. .....**Hallikar**....also known as "Mysore", the breed is considered as the best draught breed of Southern India.

34. "Swai chal" is the peculiar gait the of the ....**Kankrej..........** breed

35. ... **Kankrej** ...name comes from the name of geographical area i.e. Kankrej taluka of Banaskantha district in Gujarat.

36. .....**Krishna Valley**.....originated from black cotton soil of the watershed of the river Krishna in Karnataka and also found in border districts of Maharashtra.

37. ....**Lakhimi..........** is a dual purpose breed of cattle found in the entire state of Assam.

38. .....**Malnad Gidda........** is also known as "Gidda", "Uradana" and "Varshagandhi". "Malnad" means a hilly region and "Gidda" means small or dwarf.

39. .....**Malvi........** is named after its place of origin viz. "Malwa" region. It is also known by synonyms as "Mahadeo puri" & "Manthani". The breeding tract includes Rajgarh, Shajapur, Ratlam and Ujjain districts of Madhya Pradesh.

40. ......**Nari**..... cattle are native to Sirohi and Pali districts of Rajasthan, Sabarkantha and Banaskantha districts of Gujarat.

41. ....**Nimari**......originated from the crossing of Gir and Khillari.

42. The breeding tract of the...**Ongole**...... breed includes East Godavari, Guntur, Ongole, Nellore and Kurnool districts of Andhra Pradesh and extends all along the coast from Nellore to Vizianagaram.

43. ......**PodaThurpu**.... are medium sized cattle with compact bodies distributed in Nagarkurnool district of Telangana. Animals have white coats with brown patches or Red/brown coats with white patches.

44. The breeding tract of **.........Dangi.........** breed includes the Dangs district of Gujarat and Thane, Nasik, Ahmednagar districts of Maharashtra.

45. **Deoni.......** breed evolved from a strain descended from a mixture of **Gir, Dangi and local cattle**.

46. The breeding tract of the .....**Gaolao**.... breed includes Balaghat, Chhindwara, Seoni districts of **Madhya Pradesh**; Durg and Rajnandgaon

districts of **Chattisgarh and Wardha and Nagpur districts of Maharashtra.**

47. The dwarf breed ..... **Vechur**.... originate at a small place by the side of Vembanad lake near Vaikom in Kottayam district of South Kerala.
48. ....**Punganur ......Cattle** is one of the World,s smallest *Bos indicus* cattle originated in Punganur town in Chittoor district of Andhra Pradesh.
49. ....**Vechur** and **Punganur**.....are two miniature/dwarf cattle found in India.
50. **Thutho**....... breed of cattle is also known as "Ameshi", "Sheapi", "Chokru" and "Tseso". are available in all districts of Nagaland.
51. .....**Shweta Kapila Cattle.........** complete white coloured cattle found in North Goa and South Goa districts of Goa State.
52. ....**Purnea cattle**........... are distributed in Araria, Purnia and Katihar districts; and the Adjoining areas of Kishanganj, Supaul and Madhepura districts of Bihar.
53. ....**Sanchori**.....cattle with majority of animals are predominantly white in colour with large dewlap. It is distributed in Sanchore, Raniwara, Bhinmal, Bagoda and Chitalwana Blocks of Jalore district of Rajasthan.
54. **Masilum**....... cattle are small in size, well built, sturdy, and well adapted to the hill ecosystem of Meghalaya.
55. **Rathi.......** cattle seems to have originated from the mixture of Sahiwal, Red Sindhi, Tharparkar and Dhanni breeds apparently with a preponderance of Sahiwal blood.
56. **Red Kandhari ....**Originated from Kandhar tehsil in Nanded district of Maharashtra.
57. Karan Fries is a cross of ....**HF x Tharparkar at NDRI.**
58. Karan Swiss is a cross of ....**Brown Swiss x Sahiwal and Red Sindhi at NDRI.**
59. Jamaica hope is cross of **......80% Jersey x 15% Sahiwal x 5% HF at Jamaica.**
60. Santa Gertrudis is the cross of **........Brahman x Shorthorn at Texas.**
61. Brangus is cross of **........Brahman x Angus**

62. Guernsey is cross of **........Brown and White cattle of Brittany x Brindle cattle or Normandy**
63. **Phule Triveni** is a cross of ......HF 50% x Jersey 25% and Gir25%
64. **Hardhenu**.......comprises 62.5% exotic (from the Holstein Friesian) and 37.5% indigenous (from the Hariana and Sahiwal) components.
65. The Frieswal crossbred has ......**3/8 of Sahiwal and 5/8 of Holstein-Friesian......** inheritance.
66. Sunandini is a composite breed of cattle developed in India by crossing......... **nondescript cattle with Brown Swiss, Jersey cattle and Holstein Friesian cattle.........**
67. ....**Murrah**.... breed is also known as "Delhi", "Kundi" and "Kali".
68. ....**Murrah breed**..... typically have short and tightly curved horns.
69. ....**Murrah breed**....of buffaloes whose home is Rohtak, Hisar and Sind of Haryana, Nabha and Patiala districts of Punjab and southern parts of Delhi state, also found in Western Uttar Pradesh.
70. The home tract of .....**Nili Ravi....** buffaloes is the belt between the Sutlej and Ravi rivers of the undivided Punjab Province.
71. Due to the passage of time and with intensive crossbreeding, the two breeds converted into single breeds named .....**Nili Ravi**......
72. The name ....**Nili**.... is supposed to have been derived from the blue water of river Sutluj.
73. Nili-Ravi have walled eyes and white markings on forehead, face, muzzle, legs and tail. The most desired character of females is the possession of these white markings known as .... **"Pancha Kalyani"**...........
74. ....**Jaffarabadi..........** is one of the heaviest buffalo breeds and is a native of Saurashtra region of Gujarat around Gir forest.
75. ....In **Jaffarabadi buffalo..........** the horns are heavy, inclined to droop at each side of the neck and then turning up at point (drooping horns).
76. ....**In Surti buffalo**...Horns are flat, sickle shaped and are directed downward and backward, and then turn upward at the tip to form a hook.
77. ...**Mehsana buffalo**... evolved out of crossbreeding between the Surti and the Murrah.

78. ....**Bhadawari**...... breed is an efficient converter of coarse feed into butterfat and is known for its high butter fat content.

79. ....In **Bhadawari buffalo**...... the fat percentage in their milk ranges from 8.5% to 14%.

80. .....The **Toda buffalo.......** breed is known after its herdsmen, the tribe of the Nilgiris in the State of Tamil Nadu.

81. ....In **Toda buffalo........,** the horns are set wide apart curving inward, outward and forward forming a characteristic crescent shape or semicircle.

82. ..**In Nagpuri buffalo**....horns are long, flat, wide and thick at the base carried backwards in each side of the neck nearly up to the shoulders resembling like a pair of swords.

83. ..**In Pandharpuri buffalo**....The horns are very long, running backwards, upwards and twisted outward and touching almost the backbone. Horns are very long and extend beyond the shoulder blade, sometimes up to pin bones.

84. ....**Banni/Kundi**..... buffaloes are trained to graze on Banni grassland during night and brought to the villages in the morning for milking.

85. Horns are tightly coiled vertically with single to double coiling in ....**Banni buffalo**......

86. Home tract of .....**Bhadawari breed**.... is Agra and Etawah district of Uttar Pradesh and Gwalior district of Madhya Pradesh.

87. ....**Gojri buffaloes**...... are reared in semi-migratory/pastoral management system by Gujjar community in Pathankot, Gurdaspur, Hoshiarpur, Rupnagar and SAS Nagar (Mohali) districts of Punjab and Kangra and Chamba districts of Himachal Pradesh.

88. ....**Gojri buffaloes**...... horns are medium sized; mostly curved to form a big loop, these buffaloes are well adapted to foot hills.

89. ......**Manda buffalo**...... is distributed in Koraput, Malkangiri and Nawarangapur districts of Odisha.

90. **......Purnathadi buffalo......** name has been derived from the name of local river Purna which originates in Satpura hills and passes through Akola and Amravati districts of Vidarbha region of Maharashtra.

91. **......Purnathadi buffalo......** horns are long and tapering, may go up to the shoulder and turn upward in orientation at the end like a hook.

92. ...**Bargur breed of buffalo**...... is also known as "Malai Erumai or Malai Emmai". Malai means hills; Erumai/Emmai means buffalo. Breeding tract is Erode district of Tamil Nadu.

93. ....**Bargur breed of buffalo**....... are found in the Bargur hills in Tamil Nadu. Coat colours vary from black to light brown or brownish-black.

94. ....**Chhattisgarhi buffalo**.......... the common breeding tract is north hilly, central plains and Bastar plateau region of Chhattisgarh.

95. ......**Chilika breed**......... got its name from the name of its native tract, which is surrounding the Chilika lake in the State of Odisha.

96. This breed is known as Kalahandi in Orissa and .... **Paralakhemundi/ Peddakimedi**....... in Andhra Pradesh.

97. ....**Luit buffalo**.....Also known as "Assamese Swamp". Breeding tract includes Jorhat, Sibsagar, Lakimpur, Dibrugarh, Tinsukia, Dhemaji, Golaghat, Majuli and Biswanath districts of Assam.

98. ....**Luit (Swamp) buffaloes**.......... are mostly distributed in the upper Brahmaputra valley of Assam covering nine districts of upper Assam.

99. ... **Marathwadi buffalo**.....a distinguishing feature from Pandharpuri breed is the length of the horns, which reach only up to shoulder, unlike in Pandharpuri breed wherein they may even reach up to pin bones sometimes.

100. ....**Luit (Swamp) buffaloes**..........has chromosome number 48.

101. ....**Changthangi sheep**.......... is mainly reared by a nomadic tribe called Changpa along with Pashmina producing Changthangi goats.

102. The .....**Gaddi sheep.....** breed, also known as Bhadarwah, is native to the Kishtwar and Bhadarwah Tehsils in the Jammu region of Jammu and Kashmir, HP and Dehradun.

103. The ....**Gurej sheep........** breed is found in the **Gurez block of Bandipore district in North Kashmir**.

104. The ....**Karnah sheep**.........breed is primarily found in Karnah, a mountainous tehsil of Kupwara district in North Kashmir.

105. The ......**Poonchi**..... sheep breed, as its name suggests, is native to the Poonch and Rajouri districts of the Jammu region of Jammu and Kashmir.

106. The ....**Rampur Bushair**..... sheep breed is distributed in Shimla, Kinnaur, Nahan, Bilaspur, Solan and Lahaul and Spiti districts of Himachal Pradesh and Dehradun, Rishikesh, Chakrata and Nainital districts of Uttarakhand.

107. .....**Chokla**........ sheep also known as Chhappar and Shekhawati is native to the districts of Churu, Nagaur and Sikar in Rajasthan.

108. ..**Chokla**....sheep is perhaps the **finest carpet-wool breed**, although most Chokla wool is now being diverted to the worsted sector because of a lack of fine apparel-wool in the country.

109. .....**Chokla**........ sheep have the **face, generally devoid of wool, is reddish brown or dark brown, and the colour may extend up to the middle of the neck**; the skin is pink.

110. The ....**Magra sheep**........, also known as Bikaneri Chokhla or Chakri and formerly known as the Bikaneri, is a breed of sheep that is found in the Bikaner, Nagaur, Jaisalmer and Churu districts of Rajasthan, India.

111. .....**Marwari sheep**...are distributed in Jodhpur, Jalore, Nagaur, Pali and Barmer districts extending up to Ajmer and Udaipur districts of Rajasthan and the Jeoria region of Gujarat.

112. The ......**Marwari sheep.....** are medium size with **black face**, the colour extending to the lower part of the neck. Ears are extremely small and tubular. Both sexes are polled.

113. The .....**Nali**..... sheep is found in Ganganagar, Churu and Jhunjhunu districts of Rajasthan, southern part of Hisar and Rohtak districts of Haryana.

114. ....**Patanwadi**...... sheep is a fascinating breed found in western India, particularly in the coastal plains of Saurashtra and Kutch regions in Gujarat.

115. ....**Patanwadi**...... has **typical Roman nose with a brown face** which may be tan in a few cases. Ears are drooping, medium to large, tubular with a hairy tuft.

116. .....**Panchali**.... is a dual purpose sheep reared for milk & meat in Panchal area of Gujarat.

117. ....**Kajali sheep**......... has a white body coat with a black circle around the eyes and a black tip extending to the lower one-third of the ears.

118. ……..**Sonadi sheep**……. is found in Udaipur, Dungarpur, Banswara districts and, to some extent, Chittorgarh district of Rajasthan and also extends to northern Gujarat.

119. …..**Pugal sheep**……Pugal area of Bikaner district is its home tract. It is **distributed over Bikaner and Jaisalmer districts of Rajasthan.**

120. **…Nellore…..** is the tallest sheep with little hair except at withers, brisket, and breech.

121. **…Nellore…..** is resembling goats in appearance (Sheep-like goat), it has a long face and long ears with the body densely covered with short hair.

122. The …..**Deccani sheep……** breed derives its name from the Deccan plateau, which corresponds to its original spread across the semi-arid regions of Telangana, parts of Andhra Pradesh, Northern Karnataka, Maharashtra, and parts of Northern Tamil Nadu.

123. The ……**Hassan sheep**……. breed, as the name suggests, hails from the Hassan district of Karnataka, India. Hassan sheep are distributed in the southern parts of Karnataka.

124. The ……**Hassan sheep**……. breed has the head and neck are black or brown, the legs are long. The quality of wool is coarse rough and many times is not useful for preparing blankets.

125. ……**Bellary sheep**……. are prevalent in Chitradurga and Bellary areas. These sheep are similar to Hassan and Deccani breed in body weight and shape.

126. …..**Kenguri sheep**…..distribution is hilly tracts of Raichur district (particularly Lingasagar, Sethanaur and Gangarati taluks) of Karnataka.

127. ….**Mandya**…..sheep are bred largely in Mandya, Malavalli, and Bannur areas of Mandya district.

128. …..**Coimbatore**….. sheep are distributed in Coimbatore district and adjoining Dindigul district of Tamil Nadu.

129. ……..**Katchaikatty Black**…….sheep breed derives its name from the Katchaikatty village in Vadipatti Taluk of Madurai district, from where it originated.

130. ……**Kilakarsal sheep**……. distributed in Ramanathapuram, Madurai, Thanjavur and Ramnad districts of Tamil Nadu. Medium-sized animals.

131. The **Madras Red** ....sheep is an indigenous breed of sheep native to the northeastern parts of Tamil Nadu (Chennai, Kancheepuram, Tiruvellore, Villupuram and adjoining areas of Vellore, Cuddalore and Tiruvannamalai districts).

132. ....**Nilgiri**..... sheep is a breed of sheep found only in Nilgiris district of Tamil Nadu State in India.

133. **Ramanad-white**........ distributed in Ramnad district and adjoining areas of Tirunelveli district of Tamil Nadu.

134. **Vembur**........ sheep's body **colour is white, with irregular red and fawn patches** all over the body. Ears are medium-sized and drooping.

135. **Chevaadu**.... sheep predominantly distributing areas are Alangulam and Pappakudi panchayat union areas in Alangulam taluk; Tirunelveli taluk; Sankarankovil taluk; Nanguneri taluk of Tirunelveli district of Tamil Nadu.

136. **Macherla**....... sheep are medium to large in size, mainly in white coat colour with large brown or black patches in the body, face and legs.

137. The **Tibetan**...... sheep breed is found in northern Sikkim and Kameng district of Arunachal Pradesh.

138. **Shahabadi**...... sheep breed is found in Shahabad, Patna and Gaya districts of Bihar/Jharkhand state. The animals are of medium size and leggy.

139. **Chottanagpuri**....... sheep breed is distributed in Chotanagpur, Ranchi, Palamau, Hazaribagh, Singhbhum, Dhanbad and Santhal Parganas of Jharkhand, and Bankura district of West Bengal.

140. **Balangir**....... sheep breed is spread over north western districts of Orissa, Balangir, Sambalpur and Sundargarh.

141. **Ganjam**....... sheep are found in Koraput, Phulbani and parts of Puri districts of Orissa. The animals are medium sized with coat colour ranging from brown to dark tan.

142. **Kendrapada**.....sheep are distributed in Bhadrak, Konark and Puri district of coastal area of Orissa.

143. **Bonpala..........** this sheep breed is found in southern Sikkim.

144. **Garole....** sheep are reputed for multiple births and are found in the Sunderban area of West Bengal. These sheep are reported to have contributed prolificacy to the Booroola Merino sheep.

145. **Garole Sheep**......litter size at first lambing is two and at subsequent lambing is 2 to 3.

146. **Merino**....... sheep are famous for their **exceptionally soft and fine wool**, which was first brought to Spain in the late 19th century and has since become one of the world's most popular wool breeds.

147. The **Rambouillet.......** sheep is a breed of domestic sheep from France. It is also known as the French Merino and the Rambouillet Merino.

148. Both horned and polled ....**Dorsets**....... are all white sheep of medium size having good body length and muscle conformation to produce a desirable carcass.

149. ....**Suffolks**........ are prolific, early maturing sheep with excellent mutton carcasses.

150. The Corriedale was developed in New Zealand and Australia during the late 1800's from crossing ....**Lincoln or Leicester rams with Merino females**.

151. ......**Karakul**...... sheep are a multi-purpose breed, kept for milking, meat, pelts, and wool. As a fat-tailed breed, they have a distinctive meat.

152. **Hissardale:** Hissardale was evolved at the Government Livestock Farm, Hissar, through crossbreeding **Australian Merino** rams with **Bikaneri (Magra)** ewes and stabilizing the exotic inheritance at about 75 per cent.

153. **Harnali:** Harnali sheep is a three breed cross by 37.5% **Nali** and 62.5% exotic inheritance (**Merino and Corriedale** with equal inheritance, i.e., 31.25) developed at Lala Lajpat Rai University of Veterinary and Animal Sciences, Hisar for superior wool production.

154. The Avikalin strain has evolved from ......**Rambouillet x Malpura....** half-bred base through inter breeding and selection for wool production.

155. Avivastra is the cross of ...... **Rambouillet x Chokla** half-bred base through inter breeding and selection for wool production.

156. Avimaan is a cross of ....**Dorset and Sufflok x Malpura and Sonadi** for mutton production.

157. Bharat Merino is a cross of ....**Merino and Rambouillet** x **Chokla, Nali, Malpura and Jaiselmeri** with 75% exotic inheritance.

158. ...**Saanen**......... goats, often referred to as the "Holsteins of the dairy goat world," or queen of dairy goat.

159. ......**Angora**........ goat originated in Turkey or Asia minor, it produces a superior quality fibre called ......**mohair**.

160. ......**Changthangi and Chegu**......goat produce long hair below which under coats of delicate fibre called cashmere or pashm.

161. .....**Beetal**..... goat is found in Punjab and Haryana state but it is mostly found in Gurdaspur, Amritsar and Ferozpur districts of Punjab.

162. ......**Jamunapari goat**........ introduced near a river of Uttar Pradesh named Jamuna, since this breed is mostly known as ....**Jamunapari......** goat.

163. The ....**Barbari**......... goats get their name from Berbera, which is a coastal city located on the Indian Ocean Somalia.

164. The ....**Barbari**......... goats are popular in urban areas of Delhi, Uttar Pradesh, Gurgaon, Karnal, Panipat and Rohtak in Haryana state.

165. **Pantja**......... goat is reared for meat and milk in Udham Singh and Nainital districts of Uttarakhand and adjacent Tarai area of Uttar Pradesh.

166. The ....**.Jakhrana**.... goat breed derives its name from the Jakhrana and few surrounding villages near Behror, of Alwar district of Rajasthan where it is found in its purest form.

167. The native tract of the ......**Marwari**...... goat breed is western Rajasthan the districts of Barmer, Jaisalmer, Bikaner, Jodhpur, Jalore, Pali and Nagaur.

168. The coat colour of ....**.Sojat**.......... goat is white with brown spots on head, neck, ear and legs, however, pure white animals are also available in the field.

169. ....**Karauli**...... goats are medium to large in size and dual purpose breed, distributed in Sawai Madhopur, Kota, Bundi, and Baran districts of Rajasthan.

170. ....**.Sirohi......** is a goat breed from semi-arid region basically from Sirohi district Rajasthan.

171. The ......**Gohilwadi.......** goat is an Indian goat breed which is distributed mainly through Porbandar, Rajkot, Amreli, Bhavnagar and Junagadh districts of Gujarat state.

172. **Kahmi**...... this goat is native to Saurashtra region of Gujarat.

173. **Anjori**....... goat it is distributed in Raipur, Durg, Rajnandgaon, Kanker, Dhamtari, Mahasamund districts of Chhattisgarh state.

174. **Teressa....** goat is the first indigenous goat breed from Andaman & Nicobar Islands.

175. ...**Sumi-Ne**...... are mainly found in Zunehoboto and Tuensang districts whereas their number is very less in Kiphire, Phek and other districts of Nagaland.

176. **Black Bengal........** goats are most prolific among the Indian breeds, multiple births are common - two, three or four kids are born at a time, kidding is twice a year.

177. **Black Bengal.......** goat's skin is in great demand for high quality shoe-making, the meat is excellent and palatable.

178. **Black Bengal.......** goat's coat colour is predominantly black, brown, grey and white with soft, glossy and short hairs.

179. **Black Bengal.......** goat is best chevon breed of India.

180. **The Red Jungle fowl.......** is supposed to be the principal contributor (ancestor) for the development of modern-day poultry and is widely distributed throughout Burma, China, India, Philippines, Sumatra and Thailand.

181. American classes of chickens are ....**1) Rhode Island Red , 2) Plymouth rock, 3)New Hampshire, 4) Wyandotte**

182. Asiatic classes of chickens are ....**1) Brahma, 2) Cochin, 3) Langshan**

183. English classes of chickens are ....**1) Cornish, 2) Australorp, 3) Dorking, 4) Orpington 5) Sussex**

184. Mediterranean classes of chickens are ....**1) Leghorn, 2) Minorca, 3) Ancona, 4) Andalusian**

185. Mediterranean classes of chicken egg shell colour is ....**White**.

186. Feathered shank found in ....**Asiatic class.**

187. English classes of chicken have ....**white skin colour.**

188. Mediterranean classes of chickens have ....**white ear lobe**.

189. **White Leghorn** ....chickens are one of the most popular chickens ....**layer breeds...** due to their ability to produce the highest number of eggs per year.

190. **The Asil or Aseel** ....is an Indian breed or group of breeds of game chicken used for cock fighting and meat purposes.

191. **Kadaknath** ....also called Kali Masi or "fowl having black flesh".

192. Daothigir chickens are primarily found in the ....**Bodoland region of Assam.**

193. **Ghagus chickens ....** found in Kolar and Bangalore districts of Karnataka, Chittoor and Anantapur districts of Andhra Pradesh.

194. The name .....**Miri**......chicken is derived from the name of the tribe 'Miri' (now missing tribe) who rear these birds.

195. The name ....**"Kaunayen"**.......... is derived from two Manipuri words: "Kauna," which means "kick" or "fighting," and "yen," which translates to "hen" or "poultry."

196. The name ....**Hansli**.......... seems to have originated from the word "hans" which means swan.

197. Birds of the **Punjab Brown** breed are found in rural areas of Punjab and Haryana.

198. ....**Uttara**........... distributed in Kumaon region of Uttarakhand.

199. **Haringhata black**...... breed plays a significant role in the empowerment of tribal women through backyard poultry farming in West Bengal.

200. **Aravali Chicken.............**is a dual purpose breed, distributed in Banaskantha, Sabarkantha, Aravalli and Mahisagar districts of Gujarat State.

201. The bantam or dwarf chicken controlled by a gene (dw) is a ....**sex-linked recessive gene.**

202. The frizzle character is controlled by ....**incomplete dominant gene**.

203. The naked neck character is controlled by ....**incomplete dominant gene**.

204. Scaleless, for extreme appearance, there is an **autosomal recessive mutation** (scaleless, sc/sc).

205. The slow feathering phenomenon is a wonderful ....**sex-linked mutation.........** used for sex identification of day-old chicks.

206. **Maithili duck, Pati duck and Andamani duck.............** are registered by NBAGR.

207. **‘Katchur Anz’**............ as the only existing domestic goose species in the country registered by NBAGR.

208. **Niang Megha** and **Wak chambil......** are registered breeds of pig belonging to Meghalaya state.

209. **Zovawk**......... registered breed of pig belongs to Mizoram state.

210. **Tenyi Vo**......... registered breed of pig belongs to Nagaland state.

211. **Doom** .... registered breed of pig belongs to Assam state.

212. **Agonda Goan**......... registered breed of pig belongs to Goa state.

213. Large White Yorkshire pig Known as .... **“The Mother Breed”.**

214. **Bikaneri.........** breed of camel is one of the major camel breeds found in India.

215. The Mewari breed of camel has derived its name from the .... **Mewar area of Rajasthan...........** and is well known for milk production potential.

216. **... Kharai breed ....**is a Kutch’s unique breed of camels that can swim in seawater, known as Kharai due to its habitat and eating habits, has been identified as a separate camel by the NBAGR.

217. **... Kharai breed ......** camels have a special ability to swim in seawater and feed on saline plants and mangroves, which is how they get their name, Kharai (‘salty’ in Gujarati).

218. A small population (450 on 20th livestock census) of .....**Bactrian/ double hump**.... Camels exist in the Nubra Valley of Ladakh (Jammu and Kashmir).

219. *Camelus dromedrarius* is ......**One humped/Arabian/ Dromedary**.

220. *Camelus bactrianius* is..........**two humped/Bactrinion camel**.

221. The ....**Wessex Saddleback or Wessex Pig..........** is a breed of domestic pig originating in the West Country of England.

222. Inda has ....**eight**........... registered breeds of Horse & Pony (NBAGR).

223. The ....**Bhutia**.......horse, found in the Himalayan regions of Nepal, India, and Bhutan, is a breed of small and compact horses bearing resemblance with the Tibetan and Mongolian equine breeds.

224. A famous Sanskrit epithet about a unique feature of the ....**Marwari horse**.......... is about its neck, Mayura Greeva, which means, having a 'proud, arched neck like a peacock'.

225. Jacks are often mated with female horses (mares) to produce .... **mules**,

226. The less common hybrid of a male horse (stallion) and jenny is called .... **hinny**.

227. NBAGR-ICAR has registered ....**three**...... breeds of donkey till date, are ....**Kachchhi, Spiti and Hilary**.

228. NBAGR-ICAR has registered ....**three**.......... breeds of dog till date, are .... **Rajapalayam, Chippiparai, Mudhol Hound**

229. **Arunachali** ....yaks are the first and only recognized breed of Indian yaks

230. There are four distinct strains of Mithun and named them as .... **Arunachalee, Mizorami, Nagami and Manipuri strains.**

231. Endangered breeds of cattle are ....**Belahi, Khariar, Krishna Valley, Pulikulam.**

232. Vulnerable status of buffalo breeds are ....**Chilika and Toda....**

233. Critical status of the sheep breed is ....**Tibetan.**

234. Critical status of the goat breed is ....**Teressa.**

235. Critical status of camel breed is ......**Malvi, Mewari and Mewati**

236. Endangered pig breeds are ....**Agonda Goan and Tenyi Vo.**

237. Endangered sheep breeds ....**Karnah, Katchaikatty black and Nilgiri**

238. Endangered goat breeds ....**Chegu and Sumi-Ne**

239. Vulnerable breed of goat............. **Konkan Kanyal**

240. Only one vulnerable breed of poultry is ....**Kalasthi.**

Q2. Multiple choice questions

1. Which of the following breeds give black poultry meat ?

   a. Kadaknath b. Aseel

   c. Busra d. Cottagong

2. Which breed of buffalo has the highest fat percentage?

   a. Murrah b. Surti

   c. Bhadawari d. Mehsana

3. Which of the Indian draught cattle breed is believed to have evolved from a mixture of hilly and plain land cattle?

   a. Kenkatha b. Kherigarh

   c. **Ponwar** d. Mewati

4. As per the national cattle breeding policy improvement of well defined Indian breeds can be made through

   a. **Selective breeding** b. Cross breeding

   c. Grading up d. Top crossing

5. Pien-niu is a hybrid produced by crossing

   a. Cattle and Mithun b. Cattle and Zebra

   c. Cattle and Horse **d. Cattle and Yak**

6. Buffalo breed at risk

   a. Murrah **b. Toda**

   c. Surti d. Mehsana

7. How many varieties of Black Bengal goat are found in India?

   a. 2 b. 3

   c. 4 d. 5

8. Which of the following breeds developed for carpet wool?

   a. Nali b. Magra

   c. Avivastra **d. Avikalin**

9. Mehsana Buffalo is the cross of which following breeds?

   a. Murrah and Nagpuri
   b. Murrah and Surti
   c. Nagpuri and Suri
   d. Surti and Toda

10. Which of the following breeds has excellent heat tolerance power?

   a. Holstein Friesian
   b. Brown Swiss
   c. Ayrshire
   d. Jersey

11. Sunandini breed of cattle was evolved by crossing

   a. Local non-descriptive cows with Brown Swiss bull
   b. Sahiwal with Red Danish
   c. Red Sindhi with Red Danish
   d. none of the above

12. Many breeds have arisen due to

   a. Physiological isolation of the population
   b. Random drift
   c. Selection
   d. All of the above

13. 'Jheepra' term related to...........

   a. Sheep
   b. Goat
   c. cattle
   d. Camel

14. Best dual purpose poultry breed among following is

   a. White Leg Horn
   b. Rhode Island Rhode
   c. Black Australorp
   d. Kadaknath

15. Marwari is a breed of:

   a. Sheep
   b. Goat
   c. Camel
   d. All of the above

16. Merino of India is:

   a. Gaddi
   b. Mandya
   c. Chokla
   d. Kashmiri

17. Swiss baby's foster mother name given to a

    a. goat b. Cow

    c. Ewe d. lady

18. Swamp buffalo breed found in all of the following countries except

    a. Brazil **b. USSR**

    c. China d. Thailand

19. Marwari is not a breed of:

    a. Sheep b. Camel

    c. Horse **d. Cattle**

20. Biological name of sheep is

    a. *Bos indicus* b. *Capra hircus*

    c. *Ovis aries* d. *Bos grunniens*

21. Indigenous poultry breed include:

    a. White Leg Horn b. Minorca

    c. Daothigir d. Plymouth Rock

22. Back fat thickness is desired trait in:

    a. Poultry b. Goat

    c. Pig d. Sheep

23. Jersind is a cross of Jersey cattle with

    a. Tharparkar b. Sahiwal

    c. Gir **d. Red Sindhi**

24. National Research Centre (NRC) camel is situated at

    a. Jaipur b. Hissar

    c. Karnal **d. Bikaner**

25. Sanchori is a breed of:

    a. Cattle b. Buffalo

    c. Sheep d. Goat

26. Which native used for synthetic breed Phule Traveni ?

a. Shaiwal
b. Gir
c. Red Sidhi
d. Tharparkar

27. Exotic breed for pelt production

a. Australian Merino
b. Soviet Merino
c. Karakul
d. Rambouillet

28. Which of the following exotic breed is suitable in most part of India where climate is hot and sufficient irrigation is not available:

a. Brown Swiss
b. Ayrshire
c. Holstein Friesian
d. Jersey

29. Central Sheep and Wool Research Institute (CSWRI) is located at

a. Avikanagr
b. Bikaner
c. Karnal
d. New Delhi

30. Central Institute for Research on Buffalo (CIRB) located at

a. Hissar
b. Karnal
c. Jaipur
d. Gandhinagar

31. Central Institute for Research on Goats (CIRG) is located at

a. Makhdoom
b. Haryana
c. Karnal
d. Bareilly

32. National Bureau of Animal Genetic Resources (NBAGR) located at

a. Karnal
b. Hissar
c. Ludhiana
d. None of the above

33. Sunandini breed developed at

a. Kerala
b. Karnal
c. Gujrat
d. Maharashtra

34. Jaffarabadi buffalo belongs to

a. Punjab
b. Haryana
c. Gujrat
d. Rajasthan

35. Tharparkar cattle breed found in

    a. Gujrat
    b. Rajasthan
    c. MP
    d. UP

36. Home tract of Jamunapari goat is

    a. UP
    b. MP
    c. Rajasthan
    d. Punjab

37. Double-humped camels are popularly known as Bactrian camels found at

    a. Rajasthan
    b. Ladakh
    c. Arunachal Pradesh
    d. Nagaland

38. Which of the following goat breed produce maximum milk?

    a. Jamunapari
    b. Black Bengal
    c. Beetal
    d. Barbari

39. White strips around jaw and brisket is peculiarity of which buffalo breed?

    a. Jaffarabadi
    b. Mehsana
    c. Surti
    d. Nagpuri

40. Most prolific goat breed is ....

    a. Jamunapari
    b. Sirohi
    c. Black Bengal
    d. Beetal

41. Dwarf gene in poultry is a ....

    a. Sex-linked recessive
    b. Sex-linked dominant
    c. Dominant
    d. Incomplete dominant

42. The .... is a breed of duck. It is one of the native breeds in India, particularly found in the northeastern regions like Assam. This breed is known for its egg production capabilities.

    a. Khaki Campbell
    b. Nageshwari
    c. Indian Runner
    d. Pati

43. In hilly and costal area which exotic breed of cattle is use for crossbreeding in India.

a. **Jersey**
b. HF
c. Brown Swiss
d. Ayreshire

44. Project Directorate of Poultry is located at

a. Chennai
b. New Delhi
c. **Hyderabad**
d. Nagaland

45. Famous milch breed with half moon appearance of horn

a. Kankrej
b. Sahiwal
c. **Gir**
d. Umblacherrey

46. A native poultry breed of Gujrat with single comb

a. Naked neck
b. Frizzle fowl
c. Ghagus
d. **Ankleshwar**

47. Central duck breeding farm is located at

a. **Hessarghatta**
b. Mumbai
c. Ludhiana
d. Ahmedabad

48. Which is smallest dog breed in the world?

a. Pug
b. **Chihuahua**
c. Dalmatian
d. Pocket pom

49. Culled rabbit is called as

a. Kit
b. Bunny
c. **Roaster**
d. Weaner

50. Which of the following is a cat breed

a. Greece
b. **Bombay**
c. Athens
d. Americana

51. The status of a species is called 'Vulnerable' when population size remains

a. 0-100
b. **1000-5000**
c. 100-1000
d. 5000-10000

52. Mule is an example of

a. Cross breeding
b. Species Hybridization
c. Grading up
d. Commercial synthetic method

53. ....is a milch breed also called 'Desert Breed' due to hardiness

a. Tharparkar
b. Rathi
c. Nagori
d. Nari

54. ....is a draught type animal believed to be developed from a strain descended from the mixture of Gir, Dangi and local cattle.

a. Deoni
b. Rathi
c. Nimari
d. Nari

51. The term "jheepra" refers to a specific characteristic found in some camels, particularly the .... breed. These camels have a luxuriant growth of hair on their eyebrows, eyelids, and ears, which is what "jheepra" describes.

a. Bikaneri
b. Jaisalmeri
c. Kachchhi
d. Mewari